DISCARDED

*The Chinese Economy
Under Communism*

SOCIAL SCIENCE RESEARCH COUNCIL
Committee on the Economy of China

Simon Kuznets, *Harvard University, Chairman*
Walter Galenson, *Cornell University,*
    Director of Research
Abram Bergson, *Harvard University*
Alexander Eckstein, *University of Michigan*
Ta-Chung Liu, *Cornell University*
Sho-Chieh Tsiang, *University of Rochester*

Publications

The Economy of Mainland China, 1949–1963: A Bibliography of Materials in English, NAI-RUENN CHEN [1963]

Financing the Government Budget: Mainland China 1950–1959, GEORGE ECKLUND [1966]

Chinese Economic Statistics: A Handbook for Mainland China, NAI-RUENN CHEN [1967]

The Construction Industry in Communist China, KANG CHAO [1968]

Economic Trends in Communist China, ALEXANDER ECKSTEIN, WALTER GALENSON, AND TA-CHUNG LIU, editors [1968]

Industrial Development in Pre-Communist China, JOHN K. CHANG [1969]

Agricultural Development in China, 1368–1968, DWIGHT PERKINS [1969]

The Chinese Economy Under Communism, NAI-RUENN CHEN AND WALTER GALENSON [1969]

The Chinese Economy Under Communism

NAI-RUENN CHEN *and*
WALTER GALENSON

 LTL LAMAR TECH LIBRARY

ALDINE PUBLISHING COMPANY, CHICAGO

Copyright © 1969 by Nai-Ruenn Chen and Walter Galenson

All rights reserved. No part of this publication may be reproduced or transmitted in any form or by any means, electronic or mechanical, including photocopy, recording, or any information storage and retrieval system, without permission in writing from the publisher.

First published 1969 by
Aldine Publishing Company
529 South Wabash Avenue
Chicago, Illinois 60605

Library of Congress Catalog Card Number 69- 17556

Printed in the United States of America

Preface

Twenty years have elapsed since the Communists gained control of the Chinese mainland. During that period, a number of cataclysmic events have taken place, bound up with the Communist conception of how to move the country into modernity in the shortest possible time. After purging their opponents, real and fancied, the leaders of the new regime collectivized the huge agricultural sector and embarked upon a rapid industrialization program, which culminated in the Great Leap Forward and its aftermath of economic depression. This was followed by the Cultural Revolution, a drama that is still being played out at the present time.

It was our purpose, in writing this book, to provide a balanced summary of the economic consequences of the Chinese path to development. It was a difficult task on several counts. The huge size and the economic backwardness of the country would have made summary hazardous under even the best of circumstances. The rapidity with which basic institutions changed was another obstacle. The greatest difficulty, however, was that of securing information. During the first decade of their power, the Communists issued a relatively small amount of statistical information of dubious quality, but which nonetheless could be exploited if due care were taken. For the past ten years, however, virtually no quantitative data of any importance have come out of China.

Despite these difficulties, it is our belief that we have been able to provide a meaningful picture of the Chinese economy in transition. This achievement we owe in considerable measure to the work of a small band of scholars who have produced careful analyses of particular sectors. We relied heavily on this work, and have acknowledged it through appropriate references. There is by no means complete agreement among the specialists, even on major questions, and some of the differences are discussed in the pages that follow. On occasion, where secondary sources were inadequate we utilized primary Chinese sources.

Several colleagues were kind enough to read the original draft of this manuscript and to offer suggestions for improvement. To Professors Robert F. Dernberger of the University of Michigan, Simon Kuznets of Harvard University, and Ta-Chung Liu of Cornell University we owe a considerable debt. Mrs. Mary Carnell translated our far from legible handwriting into typed pages with speed and accuracy.

This volume is one of a series commissioned by the Committee on the Economy of China, which was appointed by the Social Science Research Council to stimulate scholarly analysis of the contemporary economic scene in China. Support for the Committee's work was provided by a grant from the Ford Foundation. Both authors have been associated with the Committee since its inception, and are grateful to the Council for financial assistance.

Particularly because much of the subject matter is controversial, it is essential to add that the authors bear sole responsibility for the final work.

N. R. C.
W. G.

Contents

ONE The Economic Heritage   1
        Agriculture
        Handicrafts
        Economic Modernization and Foreign Investment
        Characteristics of Modern Industry in China
        Foreign Trade
        Human Resources

TWO Alternative Paths to Economic Development   33
        The Resources for Development
        The Chinese Strategy for Development, The First Phase, 1952–1957
        The Great Leap Forward
        Development Policy Since the Great Leap

THREE Development of the Industrial Sector   50
        The Soviet Assistance Program
        The Growth of Industrial Output
        The Pattern of Industrial Expansion
        Handicraft Production
        The Choice of Techniques in Manufacturing
        The Location of Industry
        The Fuel and Mineral Industries
        The Electrical Power Industry
        Transportation
        The Construction Industry
        Manchuria
        Conclusions

FOUR Agriculture   87
        The Agricultural Contribution to Economic Growth
        Agricultural Production During and After the Great Leap Forward
        Agricultural Development Policy
        Current Agricultural Policy

FIVE Population and Employment   127
        Population
        Nonagricultural Employment
        Employment in Agriculture
        Unemployment
        Professional and Scientific Manpower

SIX	The Control and Allocation of Resources	143
	Economic Reorganization *Mobilization of Savings* *National Economic Planning* *The Use of Markets and Prices in Resource Allocation*	
SEVEN	Conditions of Life and Labor	166
	Pre-Communist Conditions *Aggregate Measures of Living Standards Since 1949* *Urban Living Standards, 1952–1956* *Urban Living Standards Since 1956* *Rural Living Standards* *State Welfare Benefits* *Organization of the Labor Market* *The Rural Wage System*	
EIGHT	Foreign Economic Relations	198
	Control and Organization of Foreign Trade *Trends in Foreign Trade* *The Commodity Composition of Foreign Trade* *Direction of Foreign Trade* *The Balance of International Payments* *The Chinese Foreign Aid Program*	
NINE	Prospects for the Chinese Economy	215
	Economic Growth Since 1952 *China and India Compared* *Future Policy Alternatives* *An Optimum Economic Policy for China*	
	Bibliography	230
	Index	241

List of Tables

I–1	Indexes of the Annual Average Volume of China's Foreign Trade, 1864–1936	23
I–2	Estimates of Chinese Population, 1910–1936	27
I–3	Occupational Distribution of the Chinese Population, 1933	28
II–1	Resources of China, India, and the Soviet Union at the Outset of the First Five Year Plans	35
II–2	Rates and Patterns of Investment in China, India and the Soviet Union During Their Respective Five Year Plans	38
III–1	Indexes of Chinese Industrial Output, 1949–1959	56
III–2	Indexes of Industrial Production for China, India, and the Soviet Union, for the First Seven Years of Their Respective Planning Periods	57
III–3	Conjectural Estimate of Chinese Gross Value of Industrial Production, 1957–1965	59
III–4	Industrial Output in China, by Product, 1952–1957	62
III–5	Increases in Output of Selected Chinese and Indian Industrial Products for Specified Periods	64
III–6	Output of Selected Industrial Products in China, 1957 to 1959	64
III–7	Estimated Output of Selected Industrial Products in China, 1960–1966	66
III–8	Gross Value of Handicraft Production in China, 1952–1957	67
III–9	Structure of Chinese Handicraft Production, 1954	68
III–10	Some Technological Ratios in Chinese Industry, 1952, 1957, and 1958	71
III–11	The Industrial Structure of Manufacturing Industry in China, by Province, 1952 and 1957	74
III–12	Output of Mineral Products in China, 1949 to 1959	76
III–13	Indexes of Construction Output in China, 1950 to 1958	83
IV–1	Agricultural Production Indexes for China, 1949–1957	89
IV–2	The Ratio of the Growth of Agricultural Product to the Growth of Total Product in China, 1952–1957	92
IV–3	The Ratio of the Growth of Agricultural Net Product Per Worker to the Growth of Total Net Product Per Worker in China, 1952–1957	93
IV–4	Estimates of Grain Production in China, 1949–1957	94
IV–5	Harvest Conditions and Annual Rates of Industrial Growth in China, 1952–1957	95
IV–6	The Supply of Cotton and the Production and Sale of Cotton Textile Products, 1952–1957	96
IV–7	Estimates of Grain Output in China, 1957–1967	98
IV–8	Estimates of Cotton Production in China, 1957–1966	102
IV–9	Estimates of Sown Acreage in China, 1952–1965	107
IV–10	A Comparison of Average Crop Yields in China with Selected Countries, 1952 to 1957	108

List of Tables

IV–11	Consumption of Chemical Fertilizers in China, 1952–1967	114
V–1	Estimates of the Population of Mainland China, 1964–1965	128
V–2	Employment in Chinese Industry, 1952–1957	132
V–3	Nonagricultural Employment in China, 1952 and 1957	133
V–4	The Structure of Nonagricultural Employment in China and the Soviet Union	134
V–5	Estimates of Nonagricultural Unemployment among Urban Males in China, 1949–1960	136
V–6	Professional, Technical, and Related workers as a Proportion of Total Population and of the Economically Active Male Nonagricultural Population, China and Selected Countries	141
VI–1	Pre-War and Post-War Rates of Investment in China	154
VI–2	Sources of State Budgetary Revenue in China, 1950–1959	155
VII–1	The Structure of Household Expenditures in Pre-Communist China	168
VII–2	Net Domestic Product of the Chinese Mainland, 1952–1965	169
VII–3	Estimated Personal Consumption Expenditures, 1952 to 1957	170
VII–4	Food Grain Availability of the Chinese Mainland, 1949–1965	171
VII–5	Money and Real Earnings of Chinese Workers and Employees, 1949–1965	172
VII–6	The Structure of Family Expenditures of Workers and Employees	175
VII–7	Data Relative to Clothing Availability, 1952 to 1956	177
VII–8	Family Expenditures in Pre-Communist China	180
VII–9	Relative Income and Expenditures of Workers and Employees versus Peasants in Communist China	184
VIII–1	Chinese Foreign Trade, 1950–1965	201
VIII–2	Commodity Composition of Chinese Imports, 1955–1963	204
VIII–3	Commodity Composition of Chinese Exports, 1955–1963	205
VIII–4	Direction of Chinese Foreign Trade, 1950–1965	207
VIII–5	Distribution of Chinese Imports from Non-Communist Countries, 1961–1963	208
VIII–6	Distribution of Chinese Exports to Non-Communist Countries, 1961–1964	209
VIII–7	The Balance of International Payments of China for the Period 1950 to 1964	210
VIII–8	Yearly Balance of International Payments of China, 1950 to 1964	211
VIII–9	Chinese Aid Commitments to Foreign Countries, 1953–1965	213
IX–1	Selected Economic Indicators for China, 1933–1965	218
IX–2	Selected Indicators of Economic Growth, China and India, 1951–52 to 1964–65	220
IX–3	Projected Estimates of Per Capita Grain Output in 1980	224

The Chinese Economy Under Communism

CHAPTER ONE

The Economic Heritage

The Chinese Communists inherited a largely underdeveloped economy when they came to power in 1949. The modern sector built on the periphery of China's traditional economy was small. Foreign trade, despite its contributions to economic modernization, was of only negligible importance in per capita terms. The increasingly large size of the population in relation to resources had created a serious obstacle to development. The productivity of the labor force was extremely low, leading in turn to a low level of per capita income. There was little or no saving or capital formation. The Chinese people thus were caught in a typical case of the vicious circle of poverty. Their difficulties were intensified by decades of war, which had brought on a state of hyper-inflation. As the twentieth century approached the half-way mark, the Chinese economy was disintegrating.

So that the reader may better understand the economic heritage of the Chinese Communists, this chapter presents an elaboration of this very general statement. Systematic statistics for the Chinese economy during the pre-Communist period are scanty. Where data can be found, they are usually of dubious quality. In presenting our analysis, we are forced to rely primarily on fragmentary data, including both official and scholarly estimates.

In an underdeveloped economy, it is often difficult to draw a clear line between traditional and modern components. Broadly speaking, however, a rough distinction may be made between them. By the traditional economy we mean sectors or characteristics of economic activity which are of native origin. In China's traditional economy, organization and techniques of production and distribution, as well as attitudes and ways of living and doing business, had evolved over the centuries and were not subject to swift and frequent change. By contrast, the modern economic elements were foreign in origin, usually representing a newer way of doing things and additionally characterized by constant subsequent innovation. Classified as components of the traditional Chinese economy are agriculture, handicrafts, native banks, indigenous transportation, and most trade and service activities. Large-scale factories and mines, railways, highways, and some trade and banking activities are considered modern. During the period 1931–1936, 80 per cent of China's gross national product was derived from the traditional sectors (Liu, 1946, p. 10).

Agriculture

Among the traditional economic sectors, agriculture was by far the most important. Roughly 80 per cent of the Chinese population was, and still is, dependent upon agricultural pursuits. In the 1930's agriculture accounted for approximately 60 per cent of the nation's net national product, (Ou, 1946, pp. 547–54), and 60 to 70 per cent of export trade (Li, 1951, pp. 492–506). After two decades of Communist rule, agriculture remains the dominant sector of the Chinese economy. However, the rural framework, which determined the economic fate of the peasants for centuries, has undergone fundamental changes. There were two outstanding features of this framework: (1) the minute scale of land utilization, together with laborious intensity of cultivation; and (2) an unequal distribution of land ownership.

China has an area of 9,561,000 square kilometers (not including the 36,000 square kilometers of Taiwan) (*People's Handbook for 1964*, p. 122), slightly larger than that of the United States, including Alaska and Hawaii (9,363,396 square kilometers). Limited by topography and climate, the amount of land in China suitable for cultivation is relatively small. According to one estimate, the arable land of China is not more than 15 per cent of its total land area, roughly 1.4 million square kilometers, or 350 million acres (Shen, 1951, p. 6), while the corresponding figure for the United States is 51 per cent. The total area of cultivated land was estimated by the Nationalist government in 1948 at 232.6 million acres (*Ibid.*, p. 6), roughly 10 per cent of the area of the country. These two estimates taken together suggest that some 117 million acres of arable land remain uncultivated. The accuracy of these statistics is open to question; the outstanding fact commented upon by all observers of Chinese agriculture is that the greater proportion of cultivable land in China had long been occupied and tilled, and that any remaining areas could be opened up only with a considerable amount of capital investment.

This paucity of arable land, combined with the large and fast-growing Chinese population, most of whom depend on the soil for their livelihood, has resulted in great pressure of population on the land. The cultivated area per capita declined from 0.86 acres in the middle of the seventeenth century to 0.7 acres in the 1930's and 1940's, although the cultivated acreage more than doubled in the last three hundred years (Chiu, 1951, p. 469, and Ho, 1951, pp. 6–11). Measured in terms of cultivated acreage per male farmer, there were, in the middle of the present century, approximately ten acres in China, compared with eight acres on the Indian subcontinent, 11 acres in Japan, 81 acres in the U.S.S.R., and 208 acres in the United States (Clark, 1960, p. 308). The bulk of the Chinese population lives on small areas of fertile land in river basins with little urbanization, making pressure on the land worse.

One consequence of this agricultural overpopulation was fragmentation of landholding, a problem heightened by the traditional system of dividing land more or less equally among surviving heirs. Buck's sample studies indicate that the average size of the Chinese farm during the period 1929 to 1933 was 3.76 acres, compared to 157 acres in the United States (Buck, 1956, p. 268). For some densely populated provinces, such as Chekiang, Szechwan, Yunnan, and Kweichow, the average size was only from 1.0 and 1.3 acres per holding (Lieu and Chen, 1928, pp. 181–213). Within these tiny farms was a further subdivision of fields, which were frequently scattered over an area of a half mile or more. Such an arrangement was primarily the result of an age-old practice by which peasants pooled the risk of flood and drought. Sample surveys show that there were often six to eight fields in a single farm, each, on the average, a third of a mile distant from the farmstead (Buck, 1930, p. 23). This fragmentation resulted not only in a waste of time to the farmer by requiring travel from one field to another, but also a waste of land tied up in boundaries and footpaths.

In addition, the distribution of land ownership in China was quite unequal. While it is true that large landed estates in the European sense rarely existed in China and, in fact, 15 to 20-acre farms were considered very large, the problem of land distribution was very acute.

In traditional China there were two types of land ownership: private and collective. Collective land included royal and government estates, land for military colonization, and temple and ancestral land. At the beginning of the Ch'ing Dynasty (almost four centuries ago) it was estimated that almost half the total acreage of cultivated land in China was privately owned, with the other half remaining in collective ownership (royal and government estates, 27.2 per cent; land for military colonization, 9.2 per cent; and temple and ancestral land, 13.6 per cent) (Chen, 1932, p. 11). Through social and economic developments and the weakening of dynastic control, collective ownership of land gradually disintegrated during the second half of the Ch'ing regime. According to one estimate, 92.7 per cent of the total acreage of cultivated land was in private hands in 1865 (Hsiao, 1923, pp. 336–39). The trend toward private land ownership continued into the present century. Buck's survey study, conducted during 1929–1933 for 111 *hsien* (counties) in 20 provinces, indicated that 93.3 per cent of the cultivated land belonged to private individuals (Buck, 1956, p. 193).

A feature of private land ownership in China was the concentration of land in relatively few hands. In the early 1930's, landlord families, which accounted for about 3 per cent of the farming households, owned approximately 26 per cent of the total cultivated acreage; about 68 per cent of the farm families classified as poor possessed only 22 per cent of the cultivated land; while the remaining 52 per cent of the land was owned by rich and middle peasant families.[1] Moreover, the land owned by landlords and rich peasants was usually of good quality, while that owned by middle and poor peasants was largely marginal. Thus, in a relatively rich area of the South, landlord families occupied 30 per cent of the cultivated acreage, while in the relatively poor North the corresponding estimate was 18 per cent (Wu, 1941).

Not every farmer owned the land which he operated. Over half the farm workers in traditional China were either tenants or landless laborers, and the proportion of owner-operators in the farming population tended to decline during the first half of the present century. According to data compiled by the National Agricultural Research Bureau in 1912, the first year of the Republican period, 49 per cent of the Chinese farmers owned their land outright, 28 per cent of them partly rented the land, and the remaining 23 per cent were landless tenants. Available data suggest that the degree of tenancy may have increased during the first 26 years of the Chinese Republic (Chiao, 1945, pp. 232–33).

The causes of the rising proportion of tenancy were many. War, famine, excessive taxation, exploitation by the moneylender, and declining agricultural prices were among the most important factors which had ruinous effects on small landholders. Many of them had to sell their land to make ends meet in a bad year, then becoming tenants or shifting to other occupations. Some of them even went into banditry for lack of other employment alternatives.

1. A more detailed breakdown of these data is as follows:

Distribution of Cultivated Land among Various Peasant Groups in China, 1934

Family group[a]	Average size of landholding (mou)	Per cent of total farming households	Per cent of land owned
Total	20	100	100
Landlord	1,733	3	26
Rich peasant	77	7	27
Middle peasant	23	22	25
Poor peasant	7	68	22

Source:
Wu, 1944, p. 128.

[a] Peasant families in China were customarily classified as rich, middle, and poor. The rich peasant family was one which had a relatively large amount of annual income that would usually yield a certain surplus over consumption expenditures and the cost of farm operation. The middle peasant family normally had sufficient income to meet necessary outlays, while the poor peasant family could not have met these outlays even in normal times.

In 1937, the year in which the Sino-Japanese War broke out, only 46 per cent of the Chinese farmers were occupying owners, while the proportion of part-tenants and tenants in the total farming population rose to 30 per cent and 24 per cent respectively. The degree of tenancy varied widely from region to region. Generally speaking, it was more marked in South China than in North China. In 1937, for example, 52 per cent of the farming households in Szechwan Province of Southwest China and 47 per cent of such households in Kwangtung Province of South China were tenants, while the corresponding estimates for Shantung and Hopei Provinces in North China were 10 and 11 per cent respectively (Chiao, 1945, p. 233).

The minute scale of farming and the concentration of landholding in the traditional agriculture of China together resulted in both inability and lack of incentive on the part of the Chinese people to accumulate capital. It was estimated that average capitalization per Chinese farm never exceeded $800. Nine-tenths of this sum was in land and buildings (Directorate General of Budget, Accounts, and Statistics, 1942, pp. 6–7). Lack of capital formation for productive purposes constituted the main obstacle to Chinese economic development.

Farm income was typically low. There were several factors responsible for this, the most important being the low level of labor productivity. The prevalence of small-sized farms, coupled with an abundance of farm hands, necessitated cultivation highly intensive in the use of labor. Methods of cultivation were primitive and unscientific. Farm work was performed almost entirely by human labor equipped with simple and crude implements. Except in the northeastern part of the country, power-driven machinery was not employed in farming, and even draft animals averaged less than one per farm. Irrigation was carried out largely through such laborious means as lifting the water from canals or pools by long-handled dippers or crude foot or animal-operated boosters, or in some areas from wells with hand-operated windlasses and buckets. Fertility was conserved primarily by the use of night soil and other forms of organic matter, with only limited application of chemical fertilizers.

In consequence, labor requirements in Chinese agricultural production were much greater than for the developed countries. For example, the number of man-hours required during a year to cultivate an acre of wheat was 244 in China, compared with about seven in the United States and 30 in the U.S.S.R.; one acre of cotton required 657 man-hours in China, compared with 88 man-hours in the United States and 330 man-hours in the U.S.S.R.; and the labor requirement for one acre of maize was 269 man-hours in China, compared with 24 man-hours in the United States and 69 man-hours in the U.S.S.R. (Clark, 1960, Table XXIII).

The use of labor-intensive methods of farming inevitably resulted in low output per farmer. Buck's estimates indicate that, on the basis of grain production per man-equivalent (one farmer working a full year), during 1929–1933, China produced only 1,400 kilograms, compared with 20,000 kilograms in the United States, or one-fourteenth as much (Buck, 1956, p. 15). Agricultural labor productivity in China also compared unfavorably with that of other under-developed countries. Measured in terms of Colin Clark's "Oriental Unit,"[2] real product per farmer in the wheat and millet growing area of Northern China was the lowest anywhere in the world. The analogous estimate for the most productive region of China, the Southwestern rice area, was higher than that for India and Pakistan but smaller than that for Brazil, Ceylon, Egypt, Java, Peru or the Philippines.[3]

From the standpoint of land productivity, however, Chinese performance was quite impressive. Yields per acre were by no means always lower in China than in advanced countries. In the early 1930's, for example, yields per acre of rice and wheat, the staple foods of the Chinese people, were higher in China than in the United States and the U.S.S.R., but lower than in Japan.[4] The fact that low yields per man in Chinese agriculture were associated with relatively high yields per acre is not difficult to explain. In densely populated, backward areas such as China, limited land and capital per man lead to low labor productivity, but to high yields per acre because of the high density of labor on the land.

2. The Oriental Unit, or O.U., is defined by Clark as "the quantity of goods or services exchangeable directly or indirectly for one rupee in India in 1948–49 (1960, p. 20).
3. Clark's estimate of real product per male engaged in agriculture measured in O.U. for nine poor countries was as follows:

Brazil	2,060	(1949–52)
Ceylon	1,285	(1949–52)
Egypt	1,265	(1949–52)
India	585	(1949–52)
Java	1,140	(1934–38)
Pakistan	825	(1949–52)
Peru	1,545	(1934–38)
The Philippines	1,000	(1949–52)
China		
North	250	(1929–33)
Southwest	846	(1929–33)

Clark, 1960, p. 277.
4. Data on yields of rice, wheat, barley, corn, Irish potatoes and cotton for China, Japan, India, the U.S.S.R. and the United States during the early 1930's are given in terms of kilograms per acre as follows:

Country	Rice	Wheat	Barley	Corn	Irish potatoes	Cotton
China	67	16	19	21	87	168
Japan	68	25	36	22	139	199
India	29	11	—	15	—	80
U.S.S.R.	—	10	16	15	128	188
U.S.	47	14	22	25	108	177

Sources:
Data for China were obtained from a survey of 16,334 farms in 162 localities among 150 *hsien* of 22 provinces during 1929–1933 by the Department of Agricultural Economics of the University of Nanking and were given in Buck, 1956, p. 226.
Data for other countries were taken from the *International Yearbook of Agricultural Statistics, 1930–1931,* and *1932–1933,* as given by Buck.

However, this accounts only in part for the relatively high land productivity of Chinese agriculture, since yields per acre in China were higher not only than those in advanced countries, but also in some less developed areas where the degree of labor intensity on the land was similar to that of China. As noted above, in the early 1930's, yields per acre of major crops were much higher in China than they were in India. Two explanations may be offered. First, although current investment in Chinese agriculture was very small, there was a large accumulation of man-made structures which represented an extensive and long-continued investment of labor over the centuries. These irrigation facilities, farmstead buildings, and the like in turn made possible a more efficient application of labor to the soil. Secondly, the Chinese farmer generally impressed observers by his thriftiness and industry. Although he rarely had received even a rudimentary education, year after year he cultivated the same type of land, became familiar with the same crops, used the same techniques of cultivation, and brought the same skills to bear in agricultural production. Few significant inefficiencies could be found in the allocation of the limited resources at his disposal, except perhaps for his own labor. A crude optimality was evident in Chinese agriculture, making possible relatively high yields per acre from the overcrowded soils.

Low labor productivity was by no means the only factor contributing to the low level of farm income. Several other forces were working against the Chinese farmer. First of all, there were the high rents charged by landlords. Generally speaking, three types of rental systems prevailed in China: (1) share rent; (2) crop rent; and (3) cash rent. In the system of share rent the tenant divided with the landlord in specified proportions, generally half and half, the main crops raised on the rented land. Under crop rent, the tenant paid to his landlord a fixed amount of the crop raised, or sometimes its money equivalent. Cash rent entailed payment of a fixed amount of money per unit of land. According to a survey conducted in May, 1934, by the National Agricultural Research Bureau, covering 879 *hsien* (counties) in 22 provinces, 28.1 per cent of tenant farmers paid share rent; 50.7 per cent, crop rent; and 21.2 per cent, cash rent.[5] The annual interest realized by landlords for their investment in land was 14 per cent with share rent, 13 per cent with crop rent, and 11 per cent with cash rent, compared with a normal rate of interest on long-term bank deposits in Shanghai and Peking of about 8 per cent a year (Chiu, 1951, p. 475).

In addition to paying high rentals, the tenant farmer suffered from many other landlord practices. The tenant sometimes was required to obtain credit from his landlord at high interest rates, to sell products to him, or to provide him with labor service. In some instances, landlords rented their land to the highest bidder. The tenant was frequently required to pay rent before a crop had been produced (Moyer, 1947, pp. 13–24). High rentals and questionable practices were a result of the increasing pressure of population on the land

5. *Crop Reports,* Vol. III, No. 6 (June, 1935); and Vol. IV.

and of the consequent weak bargaining power of the tenant farmer. They created a real hardship for him even in a normal crop year.

The Chinese farmer had very poor credit facilities. The need for credit, which was very pressing both for the tenant-farmer and for the peasant-owner, fell into two main categories. First, there were loans for production purposes. Since agriculture is an industry of slow turnover, the farmer, with his tiny resources, often could not provide the means to finance the interval between sowing and harvest. Then there were loans for consumption purposes, to cover, for example, unusual expenditures on funerals, weddings, or replacements after such disasters as flood or fire. Buck's sample surveys indicated that 39 per cent of Chinese farmers incurred debts, and among those in debt, 76 per cent obtained credit for consumption purposes and only 24 per cent for production purposes (Buck, 1956, p. 462).

The chief sources of farm credit were relatives and friends, landlords, rich peasants, and local merchants. The rates of interest were high. Reports received in 1933 by the National Agricultural Research Bureau from 850 *hsien* in 22 provinces indicated that the majority of farm loans (66.5 per cent) were at an annual interest rate of 20–40 per cent, and some (12.9 per cent) were more than 50 per cent.[6] The rate of interest varied from region to region. Buck found that interest rates were considerably higher in the rice areas of the Yangtze Valley and Southwestern China than in the wheat areas of North and Northwestern China (Buck, 1956, p. 462). The difference was probably due to higher farm income and the greater development of trade and communications in the rice region and lower productivity and greater risks of farming, because of uncertain crops, in the wheat region.

Third, the costs of marketing and transportation of agricultural products were unusually high in China. Chinese farms were not self-sufficient, and a fair proportion of food and materials was purchased from the outside. Rather more than half of the farm produce was marketed (Buck, 1930, pp. 356–58). An uneconomical and inefficient system of marketing and transportation tended to increase the farmer's expenditure and reduce his income. Nearly two-thirds of the farm products were sold through middlemen, whose numbers were great (Buck, 1956, p. 350, and Shen, 1951, p. 107). Adulteration by these middlemen was common in farm marketing, resulting not only in lowering the quality of the product but also in increasing its cost because of extra transportation charges for the added weight and the expense of eliminating the adulterants (Shen, 1951, p. 107).

Where farm products were sold directly to consumers, the amount of time spent on marketing was considerable. It averaged 24 days a year in both the wheat and the rice regions, but reached 72 days in the double-cropping rice area (Buck, 1956, p. 350). In addition, the local market was generally restricted geographically due to high transportation costs and tax levies on agricultural produce in transit. Costs of transportation to the local market

6. *Crop Reports,* Vol. II, No. 4 (April, 1934).

through manual means were generally higher than in the case of mechanical transport, and costs for local transportation were relatively higher than for long distances (*Ibid.,* p. 352–54). Thus it was cheaper to transport rice to Shanghai from Rangoon, Bangkok and Saigon than from Changsha in Hunan Province and Kuikiang in Kiangsi Province.[7] High transportation costs, combined with the various exactions on the way from the inland farms to the coast, effectively barred native rice from the Shanghai market.

Finally, the land tax and its surcharges were relatively high and tended to increase over time. The average amount of taxes per acre of farm land in China in the early 1930's was three to four times as much as in the United States. Over the years 1929–32, taxes per farm acre averaged $1.79 in China, and only 46¢ in the U.S. (Buck, 1956, pp. 16 and 326). In addition, farmers in China paid surtaxes on items ranging from farm implements to pigs to night soil.

It is thus clear that not only low labor productivity but also the economic pressures of high rents, excessive credit charges, large transportation and marketing costs, and burdensome taxes, had kept the small Chinese farmer at a very low level of income, which in turn made it difficult for him to abstain from the consumption necessary for savings and capital accumulation. Buck's study showed that only one-fifth of the farmers reported savings of any kind (*Ibid.,* p. 466). Among those who did save, there was little economic incentive to channel their savings into productive uses. As land became scarce and labor abundant, farmers had little interest in introducing labor-saving improvements. This was especially true for the tenant, because whatever improvements he made on the land through investment of either labor or capital belonged to the landlord and not to him.

In consequence, any impetus toward the accumulation of agricultural capital had to come from the landlord. The fact that poverty prevailed among the peasants did not imply a shortage of savings in the agricultural sector. The concentration of land holdings and the pattern of the landlord-tenant system resulted in a great inequality in the distribution of agricultural income, with a sizable amount of savings by the landlord class. But the landlord was not interested in improving the land. The most parasitic and oppressive of the landlords were the absentee owners. According to a field study conducted in 1941, 27.4 per cent of the landlords in China were absentees.[8] Completely detached from the land and its cultivation, they usually lived in the city, engaging in a business or profession. The income they derived from the land was often spent on luxurious living. While they tried to squeeze as much as they could from the output of their tenants, the landlords put little into the land in return.

7. According to a study made by the National Tariff Commission in 1932, actual transport costs for rice per *picul* from various localities to Shanghai were as follows (in Chinese dollars): from Rangoon, $0.525; from Bangkok, $0.516; from Saigon, $0.452; from Changsha, $0.952; from Kuikiang, $0.644; and from Wuhu, $0.433.

8. National Agricultural Promotion Commission, *Report,* Special Series No. II, 1942, pp. 10–11.

In sum, the traditional framework of rural China limited and discouraged capital formation in both equipment and technology. This lack of productive investment kept productivity and output at a low level. The scale of farming and the pattern of landholding as it existed in pre-Communist days were the basic obstacles to China's agricultural progress.

Handicrafts

The non-agricultural sectors of the traditional economy in China included the handicraft industry, coolie and animal transport, native banks and pawnshops, and indigenous trade and services. Among these sectors, the handicraft industry was next to agriculture in importance in the value added to net national product, the amount of employment, and the contribution to export trade.[9]

The Chinese handicraft industry may be divided into three broad classes: (1) handicrafts subsidiary to agriculture; (2) individual shops; and (3) workshops. In terms of contribution to the gross value of handicraft output, handicraft workshops were the most important as of 1949. In that year, 48 per cent of the gross value of handicraft products was created by handicraft workshops, 39 per cent by individual craftsmen, and 13 per cent by farmers.[10] Historically, handicraft activity had been a subsidiary occupation of the Chinese farmer from time immemorial. Individual handicrafts also had a long history, but gained momentum in the eighteenth century. Handicraft workshops came into being only in the eighteenth century and did not develop greatly until the second half of the nineteenth century (Sun, 1957, pp. 2–7).

We have already noted that the extreme fragmentation of land holdings and the concentration of land ownership had greatly diminished the income of the peasants. Output from the land was so small that they frequently

---

9. According to the Liu-Yeh estimates, in 1933 the net domestic product at current prices was 28.86 billion *yuan*, of which agriculture accounted for 18.76 billion, handicrafts 2.04 billion, old-fashioned transportation 1.20 billion, and peddler trade 0.96 billion. The total employment in 1933 was 259.21 million persons, of whom 204.91 million persons were engaged in agriculture; 15.74 million persons in handicraft production; 10.86 million persons in old-fashioned transportation; and 7.39 million persons in peddler trades. Liu and Yeh, 1965, pp. 66 and 69.

Data compiled by the Chinese Custom House indicate that during 1912–37, handicraft products accounted for from 25.5 to 42.1 per cent of China's export trade, with the remainder largely contributed by agriculture. Peng, 1957, p. 816 and Li, 1959, pp. 492–506

10. *Gross Value of Handicraft Output, 1949* (In millions of 1952 *yuan*)

1. Total	6,105
2. Handicraft workshops	2,868
3. Individual handicrafts	2,367
a. Urban	1,389
b. Village	978
4. Handicrafts subsidiary to agriculture	870

Sources:
Handicraft Section, 1957, p. 252 and Chao, 1957, pp. 99, 103, and 104.

could not depend solely upon it for subsistence, and had to seek supplementary income. Buck's study indicates that 14 per cent of the net income of an average farm in China was derived from subsidiary occupations, among which handicrafts were the most widely pursued (1956, pp. 298–99).

Another important reason for farmers to engage in handicrafts was the seasonal character of farm operations. The Chinese farmer was unoccupied on the land from four to six months a year, except in double-cropping areas. The period of inactivity varied with the nature and variety of crops and croppings. The winter months of November, December, January, and February provided most of the idle time (Buck, 1956, pp. 295–96 and Peng, Vol. IV, pp. 356–58). Handicrafts were a second string to the farmer's bow, supplementing agriculture when there was work neither at home nor outside.

The handicrafts which the farmer practiced during his spare hours usually were not highly specialized. They were performed in his home with his own labor and that of his family. His techniques of production were primitive; only crude and simple tools were employed. His products were generally not marketed since their quality was very low. The number and variety of the handicrafts were large; the most frequently pursued among them were hand weaving, shoe and rope making, charcoal and paper manufacturing, pottery, carpentry, and tailoring.

Individual craftsmen could be found both in rural and urban areas. In the rural areas, they were engaged primarily in such subsistence industries as smithery, carpentry, weaving, pottery-making, and tanning. But they sometimes manufactured relatively luxurious products: silk cloth, shoes, metal utensils, and furniture. These craftsmen, by catering to the needs of the peasants, contributed significantly to the economic life of the villages.

By far the largest number of individual handicraftsmen worked in cities. In 1949, for example, about 60 per cent of the gross value product of individual handicraftsmen was created in cities, with 40 per cent in the villages. Chinese urban craftsmen often attained a high degree of skill and artistic excellence, and their products became highly prized in distant markets. Well known to the West were such products as silver, jade, ivory, embroidery, silk, cloisonné, porcelain, bronze, enamel, lacquer, and furniture.

The characteristic organization of labor in the urban handicraft industry was a form of the master craftsman system, with dependent journeymen and apprentices usually related by family and neighborhood ties. Three major categories may be distinguished, despite frequent overlapping. First, the craftsman dealt directly with the consumer and worked with his own materials and equipment. Second, while the craftsman dealt directly with the consumer, he was provided with materials and was paid for his work in processing them. Finally, the craftsman was supplied with credit and raw materials by a middleman who marketed the finished products (Peng, 1957, pp. 179–80).

Handicraft workshops represented the highest stage of development in the traditional industry of China. They were a transitional form of organization, leading to large-scale capitalist factory production. They normally possessed a greater amount of capital and a higher degree of skill, with more specialization, than individual craftsmen. Mechanical power was sometimes employed in the handicraft workshops, but in the main production was carried on by hand. The master craftsman system persisted as the basic form of organization of the handicraft workshops. The number of workers hired by the handicraft workshop was, as a rule, larger than in individual handicrafts. Moreover, the workshop owner and his family members generally did not participate in the manufacturing process (Chao, 1957, p. 93).

There were inevitably differences among the various categories in structure and organization, as well as in the degree of skill and training involved and the amount of capital employed, with the handicraft workshops showing the highest degree of development and farm crafts the lowest. But there were greater similarities than differences, growing from certain difficulties which were common to all types of handicraft production in China and handicapping their development.

As with agriculture, the scarcity of capital and the abundance of labor available to the handicraft industry led to the use of labor-intensive methods and destroyed the incentive to introduce labor-saving improvements. The techniques of production were simple and primitive. Tawney's description of Chinese workshops is worth quoting:

There is little sub-division of labor or specialization of functions, and, in the majority of cases, no machinery or power. Work is heavy; craftsmanship fastidious; methods patient, laborious and slow; discipline slack or absent. Relations are human, not mechanical. There is much physical exertion and little nervous strain (Tawney, 1964, p. 114).

The productivity of labor was extremely low. In the early 1930's, the daily output of 16-count cotton yarn per factory worker was about thirty times that of a handicraftsman. In 1928, a worker in a Chinese native mine produced 50–100 kilograms of coal each day, while the daily output of his counterpart in a modern coal mine was 370 kilograms (Peng, 1957, Vol. III, pp. 648 and 702).

The other basic factors which impeded agricultural development also confronted craftsmen, namely, low income, and little or no saving or capital formation. These difficulties were aggravated by the lack of financial facilities. In order to produce and sell, the craftsman had to purchase raw materials, pay wages, and mediate between the production and the sale of the products. He could not meet his financial requirements solely out of his own small and inadequate resources. Therefore he had to borrow. But the supply of money capital was limited, and the chief lenders were middlemen, small bankers and landlords. These were expensive sources, because the loans normally carried high rates of interest or other onerous conditions.

Another handicap related to the purchase of raw materials. With neither organization nor large production scale, the handicraftsman experienced great difficulty in obtaining raw materials from a distance. Where materials were available locally, he often had to fight the competition of modern factories. He did not normally succeed in that competition. As a result, the supplies which he got were neither regular, adequate, nor reliable in quality. Furthermore, he frequently had to pay a comparatively higher price for the materials he did obtain. All this raised the price and lowered the quality of his products, and thus reduced his competitive power in the market.

He was further disadvantaged by at least two additional factors. First, being illiterate and conservative in nature, the craftsman was not always able to appreciate the need for introducing design and quality more in accord with contemporary taste. Secondly, the demand for his products was severely limited by the lack of efficient arrangements for marketing and sale, and by the lack of sufficient contact between himself and the consumer. Frequently he had to rely entirely on middlemen, an often unscrupulous breed of men. To these handicaps may be added, finally, the heavy burden of taxation imposed by local authorities.

In spite of these difficulties, the Chinese handicraft industry continued to operate, as it still does, side by side with the modern industry which was introduced in the middle of the nineteenth century. According to Ou Pao-san, handicraft production accounted in 1933 for over 90 per cent of the net value product in the lumber, transport equipment, and food processing industries; 60–80 per cent in brick, earthenware, textile clothing, leather and rubber production, and in the paper and printing industries; and over 50 per cent in the ornament and scientific instrument industries. Modern factories dominated production in the machine-building, metal processing, electrical equipment, public utilities, and chemical industries. In the aggregate, 72 per cent of the net value of manufactured products in China was produced by the handicraft industry in 1933. (Ou, 1947a, pp. 64–69 and 1947b, p. 139).

We may thus assert that traditional methods of production remained predominant in Chinese industry in the early 1930's. A more pertinent question is: had handicraft production as a whole actually declined as a result of the development of modern factories? This question is not only of importance to the study of modern Chinese economic history, but also has a bearing on Communist development strategy. Unfortunately, the question cannot yet be answered conclusively.

Available data indicate that handicraft production might have declined as much as 47 per cent from 1937 to 1949 (Chao, 1957, p. 99). But the decline was largely due to war during these twelve years. Data for pre-1937 years are not sufficient to yield a trend of the aggregate output of the handicraft industry. The conventional view is that indigenous industry in China was hampered or even undermined by its modern counterpart. This view was challenged recently by Hou, who examined statistical data on the cotton textile industry, where the competition between the indigenous and the modern producers was the keenest. He concludes that "the traditional sector existed quite well alongside the modern sector of the Chinese economy" (Hou, 1965, p. 178).

This conclusion seems plausible in view of the considerations which explain a dualistic development not only in China but also in other under-developed economies. There was considerable scope for the development of handicraft production, with competition from modern factories almost non-existent. A few cases will illustrate this point:

(1) There were a large number of products, artistic or decorative in nature, such as painting, embroidery, ivory-carving, lacquered ware, and gold and silver ware, which could not be produced with automatic machinery or on a large-scale basis.

(2) There were handicraftsmen manufacturing products to suit personal taste, as in tailoring.

(3) Farmers who engaged in handicraft work as a subsidiary occupation employed only simple implements and the raw materials at hand; their products were largely for their own consumption, with any surplus sold in the local market.

(4) There were a large number of handicraftsmen who supplied materials or component parts to modern factories. For example many Chinese families were engaged in the production of match boxes for match factories.

(5) Some handicraft workshops carried out certain processes incidental to large-scale methods of production.

(6) A large number of services followed in the wave of large-scale industries and became ancillary to them, such as repair services.

For portions of industrial production, then, handicrafts and modern factories were not competitive, and in some cases were complementary. But in many other industries there was competition. There is reason to believe that in these industries the handicraft system, in spite of its many handicaps, possessed strong resources for survival. First of all, the small scale of production, which was the inherent weakness of the handicraft industry, became an advantage when its existence was threatened by the advent of modern technology. Capital and skilled labor are prerequisites to the establishment of modern factories, but both were extremely scarce in China. A modern industry built and operated with a limited amount of capital and skilled labor could supply only a small portion of the demand; the gap was filled by handicraft production.

Moreover, labor costs were relatively low. As Eckaus and Hirschman have argued, where wage differentials exist between the traditional and the modern sectors, the average cost of production in the former may not be any higher than in the latter. Even where the unit cost of production was higher in the traditional sector than in the modern sector, this might be partially offset by the relatively low cost of distribution in the former. (Eckaus, 1955, pp. 539–635 and Hirschman, 1958, pp. 125–32). Unlike modern factories, which were concentrated in coastal areas and river ports, the handicraft industry was spread all over the country. The cost of transporting the products of modern factories into interior areas was prohibitive, and this gave a definite advantage to the traditional industry.[11] In addition, in view of its small size and close contact with the local market, the handicraft industry was better able to adjust the nature and quality of its products to local demand.

There were also instances, in a number of large cities, of handicraft workshops competing successfully with modern factories. There were over fifty thousand persons employed in the handloom workshops in Shanghai where modern textile mills clustered (Yen, 1955, p. 255). Such coexistence cannot be accounted for by a difference in the cost of distribution. Neither can it be totally explained in terms of a dual wage level. Adherence to traditional habits of consumption due to illiteracy and conservatism may have had something to do with it, but it would be wrong to say that the demonstration effect did not exist among Chinese consumers. Many new products had been accepted and had replaced traditional products. One explanation may have been the existence of a large number of low-income families whose purchasing power was so low that they could not afford the more expensive products. They continued to purchase inferior products, such as hand-made coarse cloth, instead of better but higher-priced machine-made fine cloth; their subsistence income forced them to do so (Hou, 1965, pp. 181-82). The low purchasing power of the Chinese people helped traditional industry to survive.

There are grounds for the belief that in the years prior to 1937 indigenous industry in China might not have been on the decline. This should not be taken to mean that China was doomed to a perpetual dual economy. What we are suggesting is that, under prevailing circumstances, the continued coexistence of indigenous and modern industry is understandable. These circumstances would have changed as the economy developed, permitting modern elements to spread into the traditional sector. Unfortunately, the twelve years of war after 1937 impeded economic development. Both the traditional and the modern sectors suffered greatly during these years.

Economic Modernization and Foreign Investment

The beginnings of modern industry in China can be traced back to the 1860's, when several arsenals were established by the Ch'ing government. This was a result of the recognition of the superiority of machine over handicraft production, following decades of commercial contact with the outside world and military defeats at the hands of Western industrial powers.

---

11. For example, in 1935 a bale of 16-count cotton yarn produced by a modern factory in Shanghai was priced at 168 Chinese *yuan*. When shipping the bale to certain cities in Kaingsi Province where the modern spinning industry was not developed, the distribution cost would be as follows: 17 *yuan* from Shanghai to Nanchang, 10 *yuan* from Nanchang to Linchuan, and 16 *yuan* from Nanchang to Nancheng. Thus, the selling price would be 185 *yuan* (168 *yuan* + 17 *yuan*) in Nancheng, 195 *yuan* (185 *yuan* + 10 *yuan*) in Linchuan, and 201 *yuan* (185 *yuan* + 16 *yuan*) in Nancheng. But the selling price of a bale of hand-spun yarn was 194 *yuan* in these cities. Thus, the cotton yarn from Shanghai could not compete in the local markets of Linchuan, Nancheng, and other areas further away from the main transportation lines. See Peng, 1957, Vol. III, pp. 686–687.

Apart from the state armament industry, modern factories hardly existed before 1890, when the first cotton mills appeared. Modern industry began to develop on a significant scale only after the Treaty of Shemonoseki of 1895, which permitted foreigners to operate factories in the treaty ports. Then followed the granting to the Treaty Powers of preferential or exclusive rights, known as "spheres of influence," within which the powers concerned had the privilige of leasing territory, establishing factories, mines, railroads, and banking institutions, and navigating freely on inland rivers. Modern factories established with domestic capital showed some development after the turn of the century, particularly as a result of various commercial and industrial regulations promulgated by the government to foster manufacturing.

However, due to lack of tariff protection, inadequate financial facilities, excessive internal levies, and insufficient technical knowledge, Chinese industrialists could hardly compete with foreign entrepreneurs in China who, in addition to technical and financial superiority, enjoyed the preferential treatment provided in the unequal treaties. Thus, by 1911 the domestic capital invested in modern factories amounted to only 19.3 million U.S. dollars, compared with approximately $100 million of invested foreign capital. (Y. K. Cheng, 1956, p. 27.)[12]

Industry made great strides during World War I when China was cut off from the supply of foreign consumer goods and, simultaneously, saw the demand for Chinese exports increase substantially. Between 1913 and 1920, industrial production nearly tripled (Chang, 1967, pp. 56–81). But the war-stimulated industrial boom did not take root and industrial production declined in post-war years. Due in some measure to active government participation in the economy,[13] and to rapid increase in foreign investment, including Japanese investment in Manchuria, China experienced continuous industrial expansion during the period 1923 to 1936. According to one estimate, the average annual rate of industrial growth for that period was 8.3 per cent (Chang, 1967, p. 68). The number of modern factories increased from 808 in 1920 to 2,532 in 1929, and further to 3,450 in 1933. The most notable development was the growth of the textile industry, which in 1933 had 859 factories and accounted for 42.4 per cent of total industrial production (Ou, 1947a, table after p. 64). Foreign investment in China continued to grow and reached $3,242.5 million in 1931 and $3,483.2 million in 1936 (Hou, 1965, p. 129). Foreign factories in 1933 accounted for nearly 32 per cent of industrial production and were dominant in cotton textiles, urban public utilities, tobacco and food processing, and coal mining (Y. K. Cheng, 1956, p. 211).

Between World War I and the outbreak of the Sino-Japanese War in 1937,

---

12. Remer says foreign investment in modern factories amounted to $111 million in 1914 (1933, p. 70).

13. Perhaps two-thirds of China was unified politically with the establishment of the Nanking government in 1927. The government adopted a number of economic measures to assist the growth of domestic industries. For example, the tariff rights which had been lost to Britain in 1842, and to the other Treaty Powers after subsequent military defeats, were restored in 1929. *Likin*, an internal transit or inter-port tax which had hampered the movement of commodities among regions, was abolished in 1931.

China made progress in modern transportation. Railway mileage increased from 5,800 miles in 1911 to 9,773 miles at the end of 1935. In 1911, there were practically no highways in China, and even by 1921 only 736 miles had been constructed. Highway mileage rose to 29,000 in 1930 and 59,900 at the end of 1935. Civil aviation, which began in 1929, had a total of more than 1,680,000 miles flown in 1935. Shipping increased from 87,000 tons to 720,000 tons during the period from 1913 to 1935 (*Ibid.*, p. 31).

The growth of modern industry and transportation was disrupted by the advent of the Sino-Japanese War. Manufacturing plants in the coastal areas were encouraged and assisted by the government to move inland. The equipment of 639 factories and mines was installed in various parts of the interior (Lieu, 1948, p. 15). Many new factories, largely in basic industry, were set up by the government. The number of factories in the interior provinces increased from 504 at the beginning of 1937 to 3,738 at the end of 1942 (Ministry of Economic Affairs, 1943, p. 15).

An outstanding feature of industrial development in wartime China was the relatively rapid growth of government-owned industry. In pre-war years, the government had owned very few factories and mines, but in 1942, over 42 per cent of the total industrial horsepower and 32 per cent of the labor force of 242,000 workers in the provinces controlled by the Chinese government were employed by state-operated industry (*Ibid.*, pp. 11–19). In terms of gross value of industrial production at pre-war prices, the output of government-owned industry increased by 340 per cent from 1940 to 1944, compared with an 8 per cent increase registered by private industry. The percentage share of aggregate industrial output accounted for by governmental enterprises increased from 15 per cent to 36 per cent during the same period (Y. K. Cheng, 1956, p. 265). Toward the end of the war, government enterprises began to dominate the production of iron and steel, petroleum products, and cotton yarn, while private industry still accounted for the larger part of the output of electric power, coal, chemical products and cotton piece goods.

Over-all industrial production in the inland provinces during the war years showed an upward trend until 1943, but suffered a setback during the following two years. This was due partly to the unavailability of foreign equipment and material after the Japanese blockade, and partly to the diversion of capital by some private enterprises to commodity-hoarding and speculation as a result of monetary inflation.[14] But even in the peak year of 1943,

14. Statistics of industrial production of government-controlled provinces during the war years are as follows:

Index of Industrial Production in Government-Controlled Provinces, 1938–1945
(1938 = 100)

Year	Index	Year	Index
1938	100.0	1942	302.2
1939	130.6	1943	375.6
1940	185.9	1944	351.6
1941	243.0	1945 (first half)	338.3

Source:
Cheng, 1956, p. 110.

industrial production in inland China was not high, amounting to only 147.4 million kilowatt hours of electric energy, 84,000 tons of pig iron, 10,400 tons of steel, six million tons of coal, 14.3 million gallons of liquid fuel, 127,900 bales of cotton yarn, 2,334,000 pieces of cotton goods, and a few thousand tons of basic chemicals. The aggregate value of industrial products in 1943, estimated at pre-war prices, was only about 12 per cent of the value produced in China as a whole before 1937.[15]

Industrial production in Occupied China also had its ups and downs during the war period. In the first eighteen months of the war the Japanese captured practically all the important industrial centers of China, including the Shanghai-Nanking-Hangchow triangle, the industrial districts along the North China coast, such as Tientsin, Tsingtao and Tsinan, the Wuhan industrial complex in Central China, and the Canton delta region in South China, not to mention Manchuria, which was occupied by Japan in 1932. For the first two years, Japanese efforts were geared to the restoration of the industrial capacity which had been damaged or removed to the interior. It was estimated that in the first few years, war damage in occupied China amounted to $223 million to Chinese owned industries (Han, 1945, pp. 32–39), and $800 million to foreign property.[16] Many combines and development companies were organized by the Japanese to control and wipe out Western influence in basic industries. Fragmentary data indicate that industrial capacity had recovered by 1939 and that some progress in industrial production was made during the following years.[17]

After the outbreak of the Pacific War, the economy of Occupied China, other than Manchuria, took a downturn and did not recover during the

---

15. Industrial production in government-controlled provinces as a percentage of pre-war production in China as a whole during 1940–1944 was as follows: 7.9 per cent in 1940, 10.1 per cent in 1941, 11.5 per cent in 1942, 12.2 per cent in 1943, and 11.4 per cent in 1944. Cheng, 1956, p. 265.

16. *The Trade of China* (customs reports), 1938. Vol. I, Pt. I, p. 12.

17. Very little economic information was released by the Japanese authorities in occupied China during the war period. The following table contains information on industrial growth in Shanghai and North China in selected years for which data are available.

Index of Industrial Production in Shanghai and North China, 1936–1942
(1936–1939 = 100)

Year	Shanghai	North China
1936	100.0	
1937	85.5	
1938	74.9	
1939	138.6	100.0
1940	154.8	121.0
1941	137.8	138.0
1942		148.0

Sources:
Data for Shanghai appeared in *Central Economic Monthly*, Vol. II, No. 6. Data for North China were taken from Wang, 1947.

remaining years of the Japanese occupation. In Shanghai, for example, the economy began to decline in 1941, and by 1943 industrial activity was reduced by more than half. As the Japanese forces lost ground in the Pacific in 1944, the economy of Shanghai crumbled. In the final months of Japanese rule, the production of Chinese factories in Shanghai was virtually at a standstill, while Japanese factories operated at only 25 per cent of capacity (Cheng, 1956, p. 221).

But the story of Manchurian development was different. The Japanese aim was to build a heavy industrial base there. During fourteen years of occupation, Japanese industrial investment in Manchuria amounted to $2.7 billion. As a result, Manchurian industrial production in 1944–1945 was 5.5 times greater than the pre-occupation level of 1931–1932, a spectacular growth rate of 14 per cent a year (*Ibid.*, p. 224).

Following the Japanese surrender in the fall of 1945, the Chinese government took over more than two thousand of the industrial and mining establishments which had been directly or indirectly under Japanese control in China proper, Manchuria, and Taiwan. The recovery of both state and private industries was very slow. The causes were many. There were insufficient funds and material aid for the war-ravaged industry and transportation. Damage to industry and transportation in China proper from July, 1937, to August, 1945, was estimated at $1.08 billion, and in Manchuria, including Soviet removals of machinery and equipment after the Japanese surrender, at $2 billion. (Cheng, 1956, p. 158 and Pauley, 1946). Yet in the immediate post-war years the rehabilitation funds, including the appropriations from the Chinese government (about $80 million), the supply of industrial equipment and railway repair materials obtained through foreign credit loans (about $183 million), and Japanese reparations in industrial equipment (less than $22 million), amounted to less than 10 per cent of the actual war damages and losses (Cheng, 1956, p. 158). Monetary inflation, which also hampered industrial recovery, began toward the end of the war and increased rapidly thereafter. Wholesale prices in Shanghai were 885 times higher in December, 1945, than during the first half of 1937; they rose to 5,713 times in December, 1946, to 83,800 times in December, 1947, and to 4,721,000 times in the first 18 days of August, 1948, (*Ibid.*, p. 160).

The causes of the hyper-inflation were many and need not concern us here, but the effect on industrial rehabilitation was serious. Under the inflationary pressure, the relations between cost and price were totally dislocated, making industrial planning utterly impossible. Private entrepreneurs lost the incentive to pursue productive activity; they found more profit in commodity-hoarding.

The armed conflict between the Nationalist and Communist forces immediately following the Japanese defeat intensified and finally escalated into a nation-wide civil war. The fighting, which lasted nearly four years, not only retarded economic recovery from the Sino-Japanese War but inflicted further physical damage and loss upon the economy.

Characteristics of Modern Industry in China

One salient feature of industrial modernization in China was that the process of industrialization did not start with indigenous, i.e., internally generated, forces, but was imposed by external factors, notably the Treaty Port system. This system resulted in at least three distinct characteristics of Chinese industrial development: (1) domination by foreign capital in a number of important industries; (2) uneven geographical distribution of industry; and (3) the concentration of capital investment in consumer goods industries. In addition, with the subsequent abolition of the treaty port system and the advent of the Sino-Japanese War, government ownership had become of increasing importance.

In 1936, about 42 per cent of industrial capital in China was owned by foreigners (Chao, 1957, pp. 8-9). Foreign capital was dominant in the coal, iron and steel, electric power, and textile industries. Foreign mines accounted for 55 per cent of the total coal produced in China in 1913 and for 61 per cent in 1937. All large-scale coal mines with an annual output of over one million tons were under foreign control. In 1936, foreign firms in China produced 86 per cent of the iron ore output, 80 per cent of the pig iron, and 88 per cent of the steel (Department of Industrial Statistics, 1958, p. 4). In the electric power industry, 76 per cent of the output and 68 per cent of the total generating capacity came from foreign plants in the same year (*Ibid.*, p. 38). Some 54 per cent of the spindles and 44 per cent of the looms in the textile industry were owned by foreign firms (*Ibid.*, p. 147). Other industries in which foreign-operated firms produced more than half of the total output in the 1930's included wood-working, ship-building, tanning, cigarette manufacturing, and soda and egg production (Ou, 1947b, pp. 130-33).

Domination of these important industries by foreign capital raises the question of deleterious effects upon Chinese capital. Foreign interests were not always in accord with, and in fact were frequently opposed to, Chinese interests. On the other hand, the appearance of foreign capital in China undoubtedly contributed to the formation of domestic capital. This may be inferred from the fact that many Chinese factories were established and co-existed quite well with their foreign counterparts, and that their share of the gross value product of modern industries accounted for as much as 65 per cent in 1933 (Ou, 1947b, and Hou, 1965, p. 192).

Although it is generally believed that the major sources of domestic industrial capital in China were from compradores, bureaucrats, and landlords, little research has been done on the process of capital formation. There is evidence that foreign entrepreneurs served the useful function of disseminating modern managerial knowledge and technical knowhow to the native population, giving rise to what Hou has called an "imitation effect" on the part of the Chinese entrepreneurs. This was evident in such manufacturing industries as silk reeling and spinning, flour milling, printing, and the manufacture of cigarettes, machines, and enamelware (Hou, 1965, pp. 134-36).

In addition, the creation by foreign entrepreneurs in the Treaty Ports of such social overhead capital as transportation lines and public utilities provided many external economies, and hence profitable opportunities for Chinese-owned firms. This explains in part why most of the Chinese enterprises were also concentrated in the treaty areas. Finally, the emergence of foreign investments in one industry may have led to the growth of another industry; using Hirschman's terminology, one might call this the "backward and forward linkage effect" (1958, Chapter 6). For example, the establishment of many large-scale, foreign-owned coal mines in the late nineteenth century may have had the backward linkage effect of stimulating industries which supplied materials necessary for the production of coal. At the same time, there may have been a forward linkage effect in the development of transportation which, in turn, reduced transport costs and induced further forward linkage by promoting an increase in the extent of the market for both foreign and domestic products.

Foreign capital in China was invested only in the Treaty Ports, which were either coastal or river cities. The foreign powers did not demand the right of operating factories in the inland areas, largely because transportation was inadequate and the cost of shipping was prohibitive. Chinese factories were also concentrated in the coastal and river ports because of the existence of an infrastructure there. Consequently, industries were very unevenly distributed geographically, with the coastal provinces, which comprise only 10 per cent of the total land area, producing 77 per cent of the gross value of factory output in 1949 (Data Office, 1957, pp. 67–71). Shanghai alone accounted for 54 per cent of the number of factories and of the labor force. These percentages rise to 70 and 69, respectively, if factories in Tientsin, Tsingtao and Canton are also taken into account (Tsen, 1958, p. 255). In 1943, Manchuria alone produced over 90 per cent of the iron and steel of China (Department of Industrial Statistics, 1958, p. 4). In 1936, coastal provinces accounted for 94 per cent of the generating capacity of electric power, 96 per cent of the output of electricity, 75 per cent of the metal-working factories, 88 per cent of yarn spindles, and 90 per cent of the looms (*Ibid.*, pp. 41, 138, and 150).

This pattern of location resulted not only in undue concentration of the modern economic sector, but also in considerable waste in resource utilization and marketing cost, since most factories were near neither the supply of raw materials nor the markets for their products. Steel plants in Shanghai had to secure their raw materials from thousands of miles away; pig iron came from Hopei or Manchuria, magnesium from Hunan, and iron ore from Shantung (*Ibid.*, p. 5). Manchuria and Hupeh had less than 4 per cent of the country's coal reserves but produced 55 per cent of the national coal output in 1936, while Shansi, whose coal reserves were estimated to be over 80 per cent of the national total, accounted for less than 10 per cent of production. Kiangsu Province produced only 20 per cent of the country's cotton output but possessed more than 60 per cent of the total textile capacity, while in southern Hopei, middle Shensi, and northern Honan, which were the major sources of cotton supply, the textile industry had not been developed (Tsen, 1958, pp. 255–56).

The third characteristic of industrial development in China was the imbalance between consumer and producer goods industries, with the former accounting for the bulk of the aggregate production. This was partly a consequence of foreign participation, since one of the aims of the foreign factories was to process Chinese raw materials and to sell the finished products directly to Chinese consumers. It was also due to the inability of Chinese entrepreneurs to establish heavy industrial plants; these entrepreneurs lacked the necessary financial strength, technical competence, and managerial knowledge. Prior to the Sino-Japanese War nearly 92 per cent of the industrial capital of China was invested in consumer goods industries (Chao, 1957, p. 15). In 1933, almost 85 per cent of the gross value of industrial output was created in the consumer goods industries (Liu and Yeh, 1965, pp. 142–43).

Although modern industry began in China with an armaments industry under state entrepreneurship, government participation never reached a significant scale before the Sino-Japanese War. With the movement of the Nationalist Government to the interior during the war, state ownership of industrial enterprises became of increasing importance. As already indicated, the share of the industrial production of inland provinces contributed by government-owned enterprises more than doubled from 1940 to 1944. After the war, the government took over Japanese-owned industrial enterprises with an estimated value of $1.8 billion. These became state enterprises under the control of the National Resource Commission (Y. K. Cheng, 1956, p. 266). On the eve of the Communist take-over, the Commission controlled nearly 68 per cent of all industrial capital in China, facilitating the process of nationalization.

Foreign Trade

Although commercial contact between China and the outside world can be traced as far back as the second century A.D., foreign trade as a major factor in the Chinese economy was a comparatively recent phenomenon, beginning with the opening up of trade through the Opium War with Britain in the 1840's. There were two distinct subsequent periods. The first began with the Opium War and concluded as the Sino-Japanese War began in 1937. This period was marked by a number of international conflicts, which resulted in a series of unequal treaties against China and brought forth passive trade development under the dominance of foreign economic and political forces. Despite intermittent external hostilities and internal strife, the period was relatively orderly and peaceful when compared with the second period, which included eight years of war with Japan and three years of intense civil conflict. The foreign trade data which are available for the second period are relatively incomplete and do not reflect changes in an economy totally dislocated by war and inflation. Accordingly, analysis of the volume and composition of foreign trade and its impact on development will be confined to the years prior to 1937.

Between 1864, when systematic trade records first became available, and 1936, the year before the outbreak of the Sino-Japanese War, the volume of foreign trade in China, measured in U.S. dollars, had increased about 2.5 times. Trade volume was relatively stable until the turn of the century, after which the magnitude of fluctuations became larger, as is apparent from Table I–1. Between 1900 and 1912 the trade volume doubled, with imports showing a larger increase than exports over the preceding decade. The gain in the trade volume was due partly to the opening up of Manchuria to foreign

TABLE I–1
INDICES OF THE ANNUAL AVERAGE VOLUME OF
CHINA'S FOREIGN TRADE, 1864–1936
(1913 = 100)

Period	Total	Imports	Exports
1864–1880	13.5	11.7	16.0
1881–1900	25.6	24.6	27.0
1901–1912	69.0	69.4	68.4
1913	100.0	100.0	100.0
1914–1920	109.1	101.6	119.7
1921–1930	204.0	186.8	204.4
1931–1936	139.8	152.1	122.4

Source:
Computed from Chinese Customs trade returns given in Y.K. Cheng, 1956, pp. 258–59.

trade, and partly to heavy investment in industrial, commercial, and transport enterprises by the Treaty Powers after the Shimonoseki Treaty of 1895. At the same time, foreign trade during the first decade of the century underwent a drastic change in structure. During previous decades, Chinese imports were largely offset by exports, leaving a comparatively small import excess. After the Treaty Powers had become firmly entrenched in their respective spheres of influence, machinery and equipment were imported in larger quantities, augmenting China's import balance (*Ibid.*, p. 210–211).

World War I caused import volume to be substantially reduced, while at the same time the demand for Chinese exports rose. Trade volume increased by over 50 per cent during the war period. After the war, foreign investment continued to increase, and the trade volume reached an all-time high in the 1920's, amounting to $1,150–1,776 million each year. The average annual trade volume declined from 1931 to 1936 despite moderate economic progress during these years. The decline was attributable to the world depression, the loss of Manchuria to Japan, high import prices in silver, and an upward revision of Chinese tariffs in the early 1930's (*Ibid.*, p. 212).

Although the growth of foreign trade volume in China during the seven decades up to 1936 can hardly be considered phenomenal—at peak level the total volume did not exceed $1.8 billion—changes in the composition of trade reflected industrial development. Silk and tea, the leading export commodities of the nineteenth century, were replaced by new products. Beans and bean cake, which constituted only 3 per cent of exports in 1896, occupied first place during 1928–1931, accounting for one-fifth of the total. The importance of beans and bean products declined abruptly in 1936 after the loss of Manchuria, which was the major source of supply. Taking their place as major export commodities were vegetable seeds and oil. Other commodities which gained in relative importance during the years preceding 1937 were eggs and egg products, hides and skins, mineral ores, and cotton yarn and goods (*Ibid.*, p. 34).

On the import side, the most spectacular changes involved cotton goods. Although by 1890 cotton goods and yarn had replaced opium as the leading imports, constituting 35.8 per cent of the total, their relative importance steadily declined until they reached only 1.7 per cent in 1936. This was due to the growth of the textile industry, which was also reflected in an increase in exports of cotton yarn and goods from a mere 2 per cent of total exports in 1921 to more than 8 per cent in 1931 (Li, 1951, p. 498).

The importation of other manufactured and semi-manufactured products, such as flour, sugar, tobacco, and matches, also declined as a consequence of domestic industrial growth. On the other hand, imports of industrial equipment and raw materials increased considerably. Importation of machinery was less than 1 per cent at the turn of the century and rose to 6.4 per cent by 1936. Metals and chemicals headed the list of imports in 1936, constituting 13.2 per cent and 10.8 per cent, respectively, of total imports.

At no time did foreign trade volume per capita reach four U.S. dollars. But the actual impact of foreign trade on the Chinese economy was far more important than these figures suggest. The effects of foreign trade on the growth of an underdeveloped economy are well known. It makes possible the importation of material goods (machinery, equipment, and raw and semi-finished materials) which are indispensable for industrialization. In the case of China, the effect of trade upon industry before 1937 was evidenced by the increase of imports of machinery, transport equipment, chemicals (particularly dyes), and metals. Even more important is the importation of ideas, techniques, skills, managerial knowledge, and entrepreneurship. There is no doubt that the dissemination of skills and the transmission of ideas were to a large extent responsible for the economic modernization of China.

More concretely, foreign trade was responsible for the formation of a number of industrial centers in the coastal areas. The most dramatic illustrations were the growth of Shanghai in the south and of Tientsin in the north. In the nineteenth century Shanghai was little more than a fishing town. By the 1920's it had become the leading industrial city of China, with a population of approximately one and a half million. It handled more than 40 per cent of the country's foreign trade volume. The industrial growth of Tientsin was also very rapid, concurrent with the expansion of trade volume through that city. Although there has not been a detailed study of the precise extent of the foreign trade contribution to the industrial growth of Chinese cities, examination of available statistical records reveals that a close relation, sometimes with a one-year lag, existed between the changes in the number of factories and in the volume of trade in certain coastal cities (Lieu, 1927, pp. 21–24).

The importation of certain consumer goods provided the stimulus for the rise of new industries. The outstanding example was the growth of the textile industry. We have seen that the importation of cotton goods into China exceeded that of any other item toward the end of the nineteenth century. These imported cotton goods led to the disintegration of cotton handicrafts in many areas. The superiority of the mechanical process in textile production was quickly recognized, and modern cotton mills spread. In areas where cotton handicrafts survived, techniques were frequently improved. For example, the growth of the handweaving industry in China was made possible largely through the replacement of age-old wooden hand looms by an iron type introduced from Japan (Peng, Vol. II, p. 691).

Foreign trade also provided the impetus for the growth of export industries. The existence of a large number of bean oil presses in Manchuria and the rise of steam silk filature in Shanghai are cases in point. The fur industry of Suanhwa was dependent to a great extent on the external market. The foreign demand for lace, straw braid and hairnets was responsible for the growth of workshops in Shangtung to turn out these products (Lieu, 1927, p. 23).

Finally, foreign trade introduced new consumption habits into China, yielding what Ragnar Nurkse has called the "international demonstration effect." In order to satisfy domestic demand for products theretofore unknown to Chinese consumers, a number of industries grew rapidly. Matches, cigarettes, canned food, knitted hosiery, and cement were outstanding new products. The growth of these industries was fostered in the 1920's by government regulations which exempted all machine-made articles in imitation of imported goods from part of the levies in transit (*Ibid.*, pp. 24–26).

Human Resources

The population problem has already been alluded to in our discussion of agriculture. But what was the actual size and the rate of growth of the Chinese population? Unfortunately, this question cannot be answered with any degree of precision. Prior to the Communist regime no government in China had ever undertaken a nation-wide population census in the technical sense of the term, although detailed counts were made in many localities for purposes of taxation and military conscription by a number of dynastic regimes throughout the long history of China. On the basis of fragmentary data, Ping-ti Ho has made a painstaking study of the history of population from

1368 to 1953. His conclusion gives us a useful picture of the population trend over the past centuries and is worth quoting at length:

> In conclusion, it may be guessed that China's population, which was probably at least 65,000,000 around 1400, slightly more than doubled by 1600, when it was probably about 150,000,000. During the second quarter of the seventeenth century the nation suffered severe losses in population, the exact extent of which cannot be determined. It would appear that the second half of the seventeenth century was a period of slow recovery, although the tempo of population growth was increased between 1683 and 1700. The seventeenth century as a whole probably failed to register any net gain in population. Owing to the combination of favorable economic conditions and kindly government, China's population increased from about 150,000,000 around 1700 to perhaps 313,000,000 in 1794, more than doubling in one century. Because of later growth and the lack of further economic opportunities the population reached about 430,000,000 in 1850 and the nation became increasingly impoverished. The great social upheavals of the third quarter of the nineteenth century gave China a breathing spell to make some regional economic readjustments, but the basic population-land relation in the country as a whole remained little changed. Owing to the enormous size of the nineteenth-century Chinese population, even a much lowered average rate of growth has brought it to its reported 583,000,000 by 1953. (Ho, 1959, pp. 277-78).

The great social upheaval of the third quarter of the nineteenth century which Ho refers to was the T'ai-p'ing Rebellion, believed to be the greatest civil war in world history, which resulted in a population loss of between 20 and 50 million people. Controversy has arisen as to the size of the population and the rate of its growth between the end of the T'ai-p'ing Rebellion and the beginning of the Sino-Japanese War. This was a period of comparative peace and order, and also one of greater pertinence than earlier periods to our review of economic development. With regard to size, at least a dozen estimates can be found in print for the years between 1910 and 1936, ranging from a low estimate of 368 million for 1910 published by the Ch'ing Government to a high estimate of 503 million recently prepared by Liu and Yeh. These estimates are shown in Table I-2. Among these estimates, the Liu-Yeh figure seems to be the most acceptable, not only because it is based on a wide range of data not previously available, but also because the assumptions on which it rests appear to be reasonable and sound.

Opinions also differ on the growth of population during the decades before 1937. There are two schools. One, represented by Walter F. Willcox, maintains that the population was more or less stationary from 1910 to 1937. Willcox's views were based on population changes in the highlands of southern Anhwei after the T'ai-p'ing Rebellion, as described by Hu Shih. But Liu and Yeh have questioned whether these population changes were typical of the country as a whole (1965, p. 173). The second school, which includes Buck and Liu and Yeh, suggests that the Chinese population was growing from 1912 into the 1930's. Buck's conclusion was formed on the basis of his population and agricultural survey findings during the early 1930's. Liu and Yeh accept Buck's conclusion and suggest further that "the

TABLE I–2
ESTIMATES OF CHINESE POPULATION, 1910–1936

Source	Year	Population (in thousands)
Ministry of Home Affairs, Ch'ing Government	1910	368,147
Maritime customs	1910	438,000
Walter F. Willcox	1912	381,000
China Continuation Committee	1917–18	440,925
Ch'en Ch'ang-heng	1923	443,374
Trade Handbook of the Chinese Maritime Customs	1929	444,297
League of Nations	1929	490,000
Annual Bulletin of Postal Affairs	1929	485,509
Ministry of Interior	1931	474,787
Wang Shi-ta	1932–33	429,300
Ta-chung Liu and Kung-chia Yeh	1933	503,100
Ministry of Interior	1936	479,085

Sources:
Chiao, 1945, pp. 24–25.
Willcox, 1940, p. 525.
Wang, 1935, pp. 191–266.
Liu and Yeh, 1965, pp. 178–179.

thesis of a stationary population in a more or less normal period is inconsistent with the picture of population change which Professor Willcox and Dr. Hu gave for the period after the T'ai-p'ing Rebellion" (1965, pp. 175–76).

Reliable population data are scanty, and basic employment statistics for China are even more scarce. But inadequacy of source material is not the only obstacle to the study of the occupied or employed population. The problem is complicated by the fact that China was an underdeveloped, predominantly agricultural country where labor was, as a rule, unspecialized. In an agricultural economy the labor required for farming operations varies greatly from one season to another and in consequence there is a large element of part-time or seasonal employment.

Subsidiary occupations probably contributed as much as 14 per cent of farm income in China during the early 1930's. The existence of subsidiary occupations makes the identification and enumeration of the gainfully employed labor force a difficult task. The difficulty is aggravated further by the participation of wives and children in economic activities, again frequently on a part-time or discontinuous basis. In addition, in an underdeveloped economy many occupations require only a low level of skill and of capital; people can usually move within these occupations with little sacrifice of income. Lack of labor specialization not only intensifies the problem of defining and measuring employment, but also reduces the significance and usefulness of statistics of the occupational composition of the population.

Occupational statistics may nevertheless provide a rough index of the stage of a nation's economic development and of the degree of utilization of human resources. Table I-3 presents an estimate of the occupational distribution of the Chinese population in 1933.

A few words about this table are in order. An attempt was made by Liu and Yeh to separate subsidiary non-agricultural occupations from agriculture through calculating man-labor units in agricultural and non-agricultural pursuits. When agricultural activities and subsidiary occupations were combined, the male and female working population accounted for 49 per cent and 35 per cent of their respective total populations. In aggregate terms, the

TABLE I-3
OCCUPATIONAL DISTRIBUTION OF THE CHINESE POPULATION, 1933
(in per cent)

	Male	Female	Total
Working population, age 7-64	*62.9*	*39.3*	*51.8*
Agriculture	47.2	34.0	41.0
Industry[a]	4.7	2.8	3.8
Trade	5.3	0.4	3.0
Transportation	2.9	1.5	2.2
Other non-agricultural occupations	2.8	0.6	1.8
Children under 7	19.7	19.3	19.5
Students, age 7 and over	2.7	1.5	2.2
Age 65 and over	2.5	3.7	3.0
Unemployed or idle, age 7-64	12.2	[b]	[b]
Total	100.0	100.0	100.0

Notes:
[a] Including manufacturing, utilities, mining, and construction.
[b] See text.
Source:
Liu and Yeh, 1965, p. 182.

agricultural and subsidiary employment of both males and females was 42.5 per cent of the total population in 1933 (Liu and Yeh, 1965, Tables 54 and 55).

Secondly, no estimate is made of the number of females and of the total population from seven to 64 years of age because such figures are misleading; they include not only those females who sought but could not find gainful employment, but also housewives not participating in economic activities. Finally, the lower age limit at seven seems too low for the working population. But Liu and Yeh used age seven as the dividing line between children and the working population since Buck's field study, which provided the basic statistical source for their estimates, employed this age limit.

The data confirm the low degree of development of the Chinese economy. The distribution of the labor force among major sectors of the economy was as follows: 79.1 per cent in agriculture; 7.4 per cent in industry; 5.7 per cent in trade; 4.4 per cent in transportation; and 3.4 per cent in other non-agricultural pursuits. A high proportion of the labor force engaged in agriculture is a clear indication of lack of development in an economy. In the case of China, the distributional pattern of the labor force in 1933 was very similar to that in countries with low per capita income in the 1950's.[18]

Empirical studies have suggested that with economic progress and rising incomes there is a progressive shift from primary to secondary and subsequently to tertiary categories of employment, and that in developing economies new employment in manufacturing industries tends to generate a "multiplier effect" on tertiary employment (Galenson, 1963, pp. 1–15). In the absence of occupational statistics for years other than 1933, it is not possible to provide statistical testing of these hypotheses for China. However, there is a good deal of qualitative evidence to show that during the pre-war period there was a relative increase in industrial employment in relation to agricultural employment, and that new employment in the modern manufacturing sector led to an expansion of employment in service, trade and transport activities, particularly because these activities tended to be labor-intensive as a consequence of the low level of capital and poor communications.

The economic development of a country may be judged not only by the distribution of its labor force among different sectors of economic activities, but also by the quality of its labor. We noted earlier the low labor productivity in Chinese agriculture. There has not been a comprehensive study of Chinese labor productivity in non-agricultural sectors, but a rough comparison indicates that in 1936 the value of net output per Chinese factory worker was about one-ninth that of his German or English counterpart and about one-nineteenth that of an American worker (Ou and Wang, 1946, pp. 426–34). Available evidence seems to indicate that average output per Chinese worker in the 1930's was among the lowest in the world. This was largely due to poor factor endowments. But, given factor endowments, the productivity of labor is affected by its quality, which was very low in China. Of the many things responsible for this, the most important were inadequate educational opportunities, poor health conditions, and low living standards.

Buck's sample study indicates that, in the early 1930's, about 69 per cent of the male population and about 99 per cent of the female population seven or more years of age were illiterate, with the rates of illiteracy higher in the North than in the South. Among the literate schooling was for the most part extremely brief, with males having an average of four years and females an average of less than three years (1956, pp. 373–75).

Lack of basic population and educational statistics prevents us from indicating the proportions of school-age children attending various levels of educational institutions. Fragmentary data show that the enrollment had

---

18. The average relative shares of major sectors in total employment for countries with low per capita income in the 1950's was as follows: 79.9 per cent in agriculture; 6.6 per cent in industry; 4.6 per cent in trade; 2.3 per cent in transportation; and 6.5 per cent in other sectors. See Liu and Yeh, 1965, p. 191; and Kuznets, 1957, pp. 23 and 27.

grown steadily under the Nationalist government. For example, enrollment in secondary schools rose from less than 100,000 in 1912–1913 to nearly two million in 1946–47, and enrollment in higher education increased from slightly over 40,000 to nearly 130,000 during the same period (Orleans, 1961, pp. 10, 68, and 170). But in spite of steady growth, the number of educated people in China was far from sufficient to meet the needs of the economy. Furthermore, the training provided in universities was so unbalanced that it often bore little relation to the requirements of economic development. In 1948-1949, for instance, over 55 per cent of university graduates were in liberal arts (*Ibid.*, p. 74). The resultant surplus of liberal arts graduates generally had considerable difficulty in gaining employment, while engineers, scientists, doctors, and business executives were in short supply and great demand.

In addition to the severe shortage of the professional skills needed in a modern economy, the level of skill was very low in traditional sectors. The majority of Chinese workers, particularly in agriculture, were lacking in training. The techniques which workers in the traditional sectors used in carrying out their productive activities were handed down to them through generations, and these techniques were usually inefficient and pre-scientific. For centuries the traditional technology had remained stationary. In some underdeveloped countries agricultural or handicraft extension work was established by the government to introduce new techniques. In China, however, little effort of this kind was made.

That the quality of a people is deeply affected by the condition of their health need not be elaborated. The state of a nation's health may be measured in terms of a number of basic statistics. A nation with healthy people is characterized by a low death rate, a low infant mortality rate, a low morbidity rate, and a long life expectancy.

Like population and employment statistics, health data are scarce and unreliable for China. However, some crude estimates are available. The mortality data collected by Buck indicate that in the years 1929–1931, the death rate was 27.1 per thousand for all localities surveyed, with the rate higher in the south than in the north. Registration data in Tinghsien, Hopei, yield an average annual death rate of 27.6 for the years 1932 to 1935. An average rate of 43.6 was found in Kiangyin, Kiangsu, for the years 1931 to 1934 (Buck, 1956, p. 387). From these estimates, a death rate of 30 per thousand for China might not be too far from the true average for the early 1930's, a relatively peaceful and prosperous period. This would compare unfavorably with a death rate of 24.9 for India in 1931, 18.8 for the Soviet Union in 1928, 18.2 for Japan in 1930 and 11.3 for the United States in 1930 (*Ibid.*, p. 388).

Infant mortality rates, showing death before one year of age per thousand births, also were high in China compared with those in advanced countries. During 1929–1931, China had an infant mortality rate of 161.5 for males and 155 for females, while the corresponding figures were 162 and 144 for Japan during 1921–1925, and 61 and 48 for the United States during 1929-1931 (*Ibid.*, p. 390).

These high death rates implied low life expectancy for the Chinese. At birth, the average Chinese might have looked forward to a life of about 35 years for both males and females during 1929–1931, while comparative figures for other countries were: about 27 years for both males and females in India in 1931; 42 for males and 43 for females in Japan during 1921–1925; and 59 for males and 63 for females in the United States during 1929–1931 (*Ibid.*, p. 391).

Data on Chinese morbidity rates are practically non-existent. One very rough estimate yields from 300 to 600 cases of illness per year per thousand of population, or a total of 150 million to 300 million illnesses per year in the country as a whole (Winfield, 1950, p. 107). This would mean that, on the average, between one-third and three-fifths of the entire population was sick once in the course of each year.

There existed only a very small medical force in China. In 1950, China had 41,400 modern doctors, 53,000 medical assistants, 38,000 nurses, and 16,000 midwives. This means that for every million people in China there were only 80 doctors trained in modern medicine, 70 nurses, and 30 midwives (Orleans, 1961, p. 141).

It has been generally believed that the average living standard of the Chinese people was extremely low and that a major portion of the population in China was living at the subsistence level. Lack of data on population and income does not permit one to study changes in income and consumption over time, but again, a few rough magnitudes may be useful.

One characteristic of low-income countries is the high proportion of income spent on food. According to a survey conducted for nearly 3,000 farm families in six provinces during 1922–1929, 60 per cent of consumption expenditure went for food. In some areas, food took as much as 77 per cent of the budget. The ratio of food to total expenditures was higher for families in North China than in the South, and higher for tenants than for farm owner-operators (Chiao, 1945, pp. 383–390).

Since the bulk of the family budget was spent on food, nutrition became an important measure of living standards. In terms of caloric intake, Buck's study indicates that the food energy consumption of the average Chinese was markedly above the minimum requirement.[19] This conclusion, however, requires qualification when the nutritive value of the diet is considered. The diet of the rural population was confined almost entirely to plant food. Nearly 98 per cent of the calories contained in the average Chinese diet was derived from seed products, vegetables, and fruits, as compared with about 60 per cent in the average American diet. While only 2.3 per cent of the calories in the Chinese diet were of animal origin, about 40 per cent were derived from this source in the United States. In addition, there was practically no milk in the Chinese diet. As a result, it was more bulky, lower in fat, and less digestible than the mixed diet of Westerners.

19. According to Buck, the daily standard minimum intake of energy value for an adult Chinese male was estimated to be 2,800 calories, while his actual intake during 1929–1933 averaged nearly 3,000 calories per day. 1956, p. 407.

Another characteristic of low-income countries is the relatively low proportion of income spent for housing. For an average Chinese family, housing constituted no more than 5 per cent of its budget (Chiao, 1945, pp. 383–89). By modern standards, Chinese houses were mostly of low quality, being constructed with poor materials and with inadequate ventilation and lighting.

These food and housing conditions, together with other aspects of Chinese living standards, add up to poverty. This, plus poor health and inadequate education, is more than sufficient to explain the low quality of the Chinese labor force. What is surprising is not so much the low level of productivity, but rather the ability of the Chinese peasants and workers to perform as effectively as they did. That they did is a tribute to the discipline and work-oriented traditions of Chinese society.

CHAPTER TWO

Alternative Paths to Economic Development

The immediate task of the Chinese Communist leadership, after the assumption of power in 1949, was restoration to some semblance of order of an economy badly battered by war and inflation. Despite the strains caused by the Korean War, rehabilitation had largely been accomplished by 1952. The industrial base of Manchuria was getting back into operation, railroads were repaired, and the thousands of small farms, enjoying peace for the first time in many years, began once again to produce the food and agricultural raw materials that were so desperately needed. A careful assessment of the progress of agriculture during the period reached the conclusion that "by 1952 at the latest, output was in the neighborhood of the prewar level" (Liu and Yeh, 1965, p. 53).

Given the ideological commitment of the Chinese Communist Party, it was a foregone conclusion that rapid economic growth would be an overriding goal. But the path to development had to be chosen, and here a great variety of possibilities existed. For convenience in exposition, we will consider three groups of alternative courses under the following labels: a) the Soviet model; b) the Indian model; c) the balanced growth model.

By the Soviet model we mean the economic policies adopted by the Soviet Union in 1928, particularly those associated with the name of Stalin. Prior to the adoption of the First Five Year Plan (FYP) in 1928 there had been considerable discussion within the Russian Communist Party of appropriate policy lines. Some voices were raised in favor of emphasizing the growth of agricultural productivity, but they were not numerous. The real debate was between the moderates, led by Nikolai Bukharin, who proposed joint development of agriculture and industry, and the leftists, whose chief theoretical spokesman was Evgenii Preobrazhenskii, who urged that heavy industry be accorded high priority, with agriculture to fulfill the function of providing the means of capital accumulation without itself receiving any substantial initial investment. Stalin chose the latter course and instituted collectivization of farming to facilitate the collection of agricultural commodities from a recalcitrant peasantry. For some years investment was maintained at a level of 20 to 25 per cent of gross national product, a remarkably high figure for a country at the Russian stage of development. Moreover, investment was channeled, in large measure, into heavy industry —coal, steel, metalworking, and machinery. This concentration of resources permitted an exceedingly rapid industrialization, but at a heavy price to the consumer and at the cost of a lagging and perennially troublesome agricultural sector.[1]

The Indian model represents the general approach to economic development embodied in the Indian five year plans, the third of which was completed in 1966. These are by no means entirely consistent, but taken as a whole represent an attempt to combine fairly rapid industrialization with simultaneous expansion and improvement in agriculture.[2] A substantial portion of total investment—about 20 per cent—was allocated to the agricultural sector, and considerable amounts were devoted to transportation, communications, and small scale industry. Heavy industry has certainly not been neglected, but there was not the single-minded concentration on building up this sector of the economy characteristic of Soviet policy.

The balanced growth model may be thought of as covering a broader range of experience and representing the expansion of all sectors of the economy, including agriculture, with the guideline for investment being maximization of output as measured by market price relationships. No special priority is accorded heavy industry; if, given the relative costs of labor and capital, as well as import and export possibilities, textile mills are more profitable than steel mills, then the textile industry will be expanded, not the steel industry. This strategy, while not necessarily incompatible with central planning, relies on the availability of market price indicators to determine profitability. Many countries have pursued a policy of this character without conscious knowledge that a particular course of development had been chosen.

These models are actually general categories which tend to shade into

1. It has been estimated that the real wage level of the urban population *fell* by about 25 per cent from 1928 to 1940. Chapman, 1962, p. 166.
2. The Third FYP summarizes the approach as follows: "In the scheme of development, the first priority necessarily belongs to agriculture; and agricultural production has to be increased to the highest levels feasible ... The growth of agriculture and the development of human resources alike hinge upon the advance made by industry ... Agriculture and industry must be regarded as integral parts of the same process of development. Through planned development, therefore, the growth of industry has to be speeded and economic progress accelerated. In particular, heavy industries and machine-making industries have to be developed ..." Government of India, 1961, pp. 6–7.

one another, and one might be hard put to quantify them with any degree of precision. Yet they are useful in providing a rough framework for analyzing Chinese development since 1949.

The Resources for Development

In choosing their development strategy, the Chinese planners had to take into account a pattern of land, labor, and capital resources that differed greatly from that confronting the Soviet and Indian authorities at the outset of their planning periods. Some of these differences are high-lighted in Table II-1.

TABLE II-1
RESOURCES OF CHINA, INDIA, AND THE SOVIET UNION
AT THE OUTSET OF THE FIRST FIVE YEAR PLANS

	China 1952	India 1951	Soviet Union 1928
Population (millions)	567	357	150
Crop area per head of rural population (acres)	0.7	1.1	2.3
Per capita output of grain (kg.)	272	281	566
Railroads (km. per million population)	43	160	513
Electric power generating capacity (mill. kw.)	2.0	3.0	1.9
Industrial production[a]:			
Coal (millions of metric tons)	63.5	34.3	35.5
Steel (millions of metric tons)	1.3	1.1	4.3
Cement (millions of metric tons)	2.9	3.2	1.8
Cotton cloth (millions of meters)	4,158	4,221	2,678

Note:
[a]Industrial production data are shown in absolute rather than per capita terms.
Sources:
Government of India, 1953 and 1958.
U.S.S.R. Central Statistical Office, 1957b.
Li, 1959.

The Chinese were clearly at a considerable initial disadvantage compared with the Soviet Union. The latter had a better endowment of arable land, and its grain output per capita was twice that of China. The Russian transportation network was better developed. While the Soviet Union of 1928 was by no means an advanced nation economically, it was considerably ahead of the China of 1952. In its educational and technical training levels, particularly important assets for development, the Soviet Union was substantially more advanced.

The difference between the initial positions of China and India was not as clear cut. The two countries were very much alike in land resources and per capita grain output. The Indian infra-structure was better developed than the Chinese, but the per capita availability of basic commodities was much the same. However, India appears to have had an initial edge in education, measured both in school attendance and output of professionals (See Malenbaum, 1959, p. 297). And India inherited a going civil service from the British *raj*, while the Chinese Communists had only the party apparatus from Yenan.

Certainly China (and India as well) faced considerably greater handicaps than the Soviet Union in overcoming the inertia of backwardness. Its peasants were living close to the margin of subsistence, and little food was left over to feed the thousands of workers required by new industrial enterprises. While the Soviet government purchased nine million tons of grain from its peasants in 1928, and the Chinese government ten million tons in 1952, in Russia, about twice as much was left for each person in the country-side. Moreover, the Chinese government had to feed with what it bought an urban population two and a half times as large as that of the Soviet Union (Yeh, 1967, p. 347). This crucial difference was in large measure responsible for the different results of the early development experiences of the two nations. Stalin was able to exact a continuous stream of saving from the farm sector to finance his industrialization program, but the Chinese eventually ran up against the limiting factor of inadequate food supplies.

If India had followed the Chinese development scheme, it would have run into similar difficulties, for the situation of the Indian peasant was worse, if anything, than that of the Chinese peasant. But this was never a temptation, since to impose a high rate of saving on a subsistence economy requires a degree of governmental authoritarianism that India did not have. The Russian agricultural collectivization of 1929–1930, which was undertaken in order to facilitate the collection of grain by the government, involved mass deportation and starvation of the peasantry. The Chinese collectivization program, culminating in the formation of the communes, had largely the same end in view, though it was accomplished without a comparable recourse to force. India, by contrast, relied more upon an effort to raise farm productivity through investment of additional resources than on spectacular structural changes, although there was some land reform designed to distribute available land more equitably among individual farmers at the expense of large landlords. As the first Indian FYP put it: "For the immediate five year period, agriculture, including irrigation and power, must have the topmost priority ... it is clear that without a substantial increase in the production of food and of the raw materials needed for industry, it will be impossible to sustain a higher tempo of development in other sectors" (Government of India, 1953, p. 20).

The choice of development strategies among the three countries was undoubtedly influenced greatly by their relative situations with respect to population. The Chinese had at their disposal a huge mass of manpower, dwarfing even the manpower resources of India and exceeding those of 1928 Russia almost fourfold. Moreover, the rate of population increase in

China was more rapid than that of India or Russia. As nearly as anyone can determine, the population of China was increasing at about 2 per cent per annum in 1952, adding about 12 million persons a year (Liu and Yeh, 1965, p. 102). The Soviet rate of natural increase in 1928 was about 2.1 per cent, but the turmoil created by collectivization of agriculture resulted in a net increase of only 1.3 million persons a year from 1929 to 1937, rather than the three million a year who would have been added to the population had the 1928 demographic situation continued (U.S. Bureau of the Census, 1964, p. 3). During the first Indian FYP population appears to have been growing at the slightly lower rate of 1.9 per cent per annum. (United Nations, 1966a).

China is the case *par excellence* of what W. Arthur Lewis has termed economic development with unlimited supplies of labor. It had considerable underemployment among its rural population, and both underemployment and unemployment in the cities. Estimates of the precise magnitudes involved vary greatly. One estimate puts the number of unemployed males in 1952 as high as 25 million, of whom seven million were in urban areas (Liu and Yeh, 1965, p. 102). Somewhat lower figures are given in another source (Hou, 1968, p. 369). No one has even attempted to estimate female unemployment, due to the lack of information on labor force participation rates among women in China.

India was in a similar situation. The Indian planners were unable to provide any estimate of unemployment in 1951, but they noted: "The problem of unemployment and underemployment in urban areas is equally acute [as in rural areas]. Owing to increasing pressure on the land, a large number of people move to the towns and cities to seek employment. There is thus a keen competition for unskilled jobs in the factories and in domestic service" (Government of India, 1953, p. 252).

The Soviet situation in 1928 was quite different. There was no large pool of surplus manpower in the cities that could be tapped for the industrialization drive. Although the Soviet Union was more urbanized at the time than either China or India a quarter of a century later (the proportion of the population in urban areas was 19 per cent for the USSR, 12 per cent for China, and 17 per cent for India), the relatively higher level of Soviet development meant that fewer of its city dwellers were engaged in low-productivity tertiary employment, and thus fewer of them could be moved into industrial employment without an impact on services. Soviet urban unemployment on October 1, 1928, was said to be 1.365 million, on the basis of registration at labor exchanges, but Soviet authorities claimed these people had been absorbed into the labor force by 1930 (U.S.S.R. Central Statistical Bureau, 1957a, p. 247). The Russians found it necessary to move millions of people from country to city during the 1930's. Their urban population rose by 18 million, or 62 per cent, from 1929 to 1937; the nation's total population rose by only 7 per cent during the same period. The Chinese and Indian governments, on the other hand, made strenuous efforts to arrest the cityward drift of farm dwellers; the former on occasion even resorted to deportation of urban workers back to their villages of origin (Hughes and Luard, 1959, p. 118). The Russians used organized recruitment from the collective farms to man their factories, while the Chinese and the Indians were able to satisfy their manufacturing needs—and more—from labor resources already in the cities at the inception of their industrialization periods.

The Chinese Strategy for Development: The First Phase, 1952–1957

Given the unpreparedness of the Chinese Communists to face on their own the problems posed by industrialization, it was almost inevitable that they should choose the Soviet model as a pattern for their development efforts. Certain modifications proved necessary because of their particular resource endowments. Moreover, as they gained experience and confidence, they introduced variations of their own. Their initial strategy, therefore, though Stalinist in conception and spirit, had certain unique characteristics.

The pattern of investment was strikingly modeled upon the Soviet pattern. The investment rate during the first Chinese FYP seems to have been about the same as that of the Soviet Union, despite the much lower Chinese per capita income level (see Table II–2). India, by comparison, maintained a lower rate of investment, one more in line with what is ordinarily characteristic of an underdeveloped nation.

Some modification of this statement may be required by the fact that Chinese producer goods were probably overvalued relative to consumer goods. The Liu-Yeh estimates show that in 1933 prices (which were more or

TABLE II–2
RATES AND PATTERNS OF INVESTMENT IN CHINA, INDIA, AND THE SOVIET UNION DURING THEIR RESPECTIVE FIVE YEAR PLANS
(in per cent)

	China[a]	India[b]	Soviet Union[a]
Ratio of gross investment to gross domestic product	20–25	10.5	20–25
Sectoral distribution of investment:			
Industry	47.9	25.0	40.9
Agriculture	14.9	26.5	19.2
Other	37.2	48.5	39.9

Notes:
[a] For China and the Soviet Union, the sectoral data relate only to public capital investment.
[b] For India, the sectoral data include both public and private investment.

Sources:
China: Yeh, 1967, pp. 330 and 334.
India: Malenbaum, 1959, pp. 287 and 300.
Soviet Union: Yeh, 1967, p. 4 and Bergson, 1964, p. 308.

less free market prices), the ratio of gross capital investment to gross domestic expenditure in 1957 was about 15 per cent, whereas it was over 25 per cent in 1952 prices (Liu and Yeh, 1965, p. 68). Proper revaluation of the Chinese data might well bring the investment ratio closer to that of India. Soviet producer goods tended to be similarly overvalued, though probably not as much as the Chinese. Bergson has estimated that in terms of factor cost, gross investment constituted 25 per cent of Soviet GNP in 1928 and 27 per cent in 1937 (1961, pp. 145 and 154).

The sectoral distribution also indicates that the Chinese were determined to outdo the Russians in speed of industrialization. Almost half of the Chinese First FYP investment went to industry (which includes mining and electric power, as well as manufacturing), considerably more than the first Soviet FYP allocated to this sector. Agricultural investment, on the other hand, was substantially less in proportion to total investment than in the Soviet case, though both countries were well below India in this respect. The higher Soviet share allotted to agriculture was probably the consequence not so much of a greater Soviet belief in the importance of agriculture as of the necessity to replace by machinery the draught animals slaughtered during the course of the 1930 collectivization, and the need to mechanize work being vacated by the farm labor diverted to the factories. The Chinese, on the other hand, paid more attention to irrigation and water control, partly out of a desire to employ productively their underemployed farm labor, and partly also because of the nature of their crops. Neither country, at the early stages of their development, gave much heed to the possibility of raising yields through the application of chemical fertilizers, pesticides, and improved seed.

The investment figures enable us to delineate more precisely the nature of the development models described earlier. The Soviet model implies something like a 20 to 25 per cent investment rate, with 40 per cent and more of total investment allocated to the industrial sector. The Indian model implies a lower investment rate (10 to 20 per cent) together with substantial investment in agriculture and social overhead, though industry still received a respectable slice. The balanced growth model does not imply any particular rate of investment, nor any precise distribution of investment among sectors. Everything would depend on factor proportions, resources, and foreign trade possibilities in any given case.

Another way of making these distinctions more meaningful is by looking at the allocation of investment funds within the industrial sector. The Soviet model stressed so-called heavy industry—mining, basic iron and steel, machine-building and metal-working, and non-ferrous metals. Some 82 per cent of total industrial investment during the first Soviet FYP went into heavy industry, leaving only 18 per cent for the light (consumer goods) industries (Kaplan, 1953, p. 66). Heavy industry investment for the first Chinese FYP was about 85 per cent (Yeh, 1967, pp. 336–37). A similar calculation is not available for the first Indian FYP, but the allocation provided by the second FYP, which assigned a higher priority than the first FYP to heavy industry, does not seem to have begun to approach the levels of the Chinese and Soviet plans.[3]

The Chinese departed from the Soviet model, even in this first phase of Chinese planning, in a number of important respects. They are:

1. *Handicrafts.* The Russian planners gave short shrift to small workshops and the traditional handicrafts. Artisans were herded into cooperatives, whose share of total output declined rapidly.[4] Employment in small-scale enterprises fell by almost two-thirds during the first FYP (Yeh, 1967, p. 339). The stress was entirely on modern, large-scale industry; independent artisans were looked upon as remnants of bourgeois society. The absence of large-scale underemployment once industrialization got under way in the Soviet Union made it possible to reduce the role of the traditional handicrafts to insignificance.[5]

In China (and India as well) the traditional handicrafts retained their significance in both output and employment. One estimate put non-agricultural Chinese handicraft employment at 10 million in 1957, compared with only 5.9 million in factories and mining (Hou, 1968, pp. 356–57). Employment in traditional transportation exceeded that in modern transportation by a wide margin: 4.8 million to 1.9 million (*Ibid.*) During the initial stages of Indian planning, small scale enterprises contributed as much to total output as the large factories but, because of their lower productivity, employed four times as many workers (Malenbaum, 1962, p. 231).

The continuation of handicraft industries presented no ideological problem in India because of the Gandhian heritage of faith in the self-sufficient village economy. Sometimes it seemed as though the problem in that country was to prevent excessive reliance on the handicraft sector to solve the problem of unemployment.[6] The Chinese, however, had reservations about the ultimate value of small-scale industry. They visualized its transformation into cooperative workshops and eventually into full-fledged modern industrial enterprises. Nevertheless, they had not only to retain the traditional workshops during the early plan years, but even to permit their expansion. The value of handicraft output almost doubled from 1952 to 1957 (Schran, 1964a, p. 159). Handicraft employment also rose somewhat (Liu and Yeh, 1965, p. 196).

3. One estimate assigns 65 per cent of industrial investment under the second Indian FYP to basic investment goods, and 35 per cent to consumer goods. Malenbaum, 1962, p. 87.
4. Cooperative enterprises and private producers accounted for 30.6 per cent of gross industrial output in 1923, and 8.2 per cent in 1937. Nove, 1961, p. 28.
5. 1959 is the last year for which employment in producer cooperatives is shown separately. The total was 1.4 million out of a civilian labor force of 108 million. Joint Economic Committee, 1965, p. 65.
6. "... the rationale for choosing cottage industries as the principal answer to the problem of residual unemployment was exceedingly tortured. For their expansion had to be engineered at the expense of more efficient, lower-unit-cost competitors ... The cottage industry program created expectations of continuing protection among the artisans it favored and thereby spawned pressure groups dedicated to the perpetuation of pockets of economic insufficiency." Lewis, 1964, p. 64.

There was a major difference in handicraft policy between the two countries. India allotted substantial investment funds to this sector, while China grudgingly tolerated its existence and did little to encourage it with new infusions of capital.

2. *Agriculture.* The Chinese Communists, while just as interested as the Russians in extracting as much food and industrial raw materials from the countryside as possible, showed greater expertise than the Russians in handling the peasants. The Bolshevik leaders had been mainly city people, but the Chinese were tempered by their enforced residence in Kiangsi and Yenan.

When the civil war ended in Russia, the Communist government was cautious about taking any steps that might alienate the peasants. They were permitted to retain possession of the land, including that expropriated from Czarist landowners. By 1925, prewar levels of farm output had been regained. Beginning in 1928, however, grain collection became difficult, since the peasants were unwilling to sell at the prices set by the government. Stalin's answer was collectivization, with its attendant disruption of rural life.

The Chinese started by redistributing 46 million hectares of land from wealthy to poor peasants. This helped them to gain the goodwill of the peasantry, and thus to consolidate their control of the country. Once this was done, they began to urge the peasants to join mutual aid teams, in which seven to eight households combined their efforts for peak season work. By 1954, about 60 per cent of all peasant households had entered into this arrangement. During 1955–1957, the teams were consolidated first into loose cooperatives, and then into full-fledged collectives, in which the land was owned in common and each member was paid on the basis of the number of "labor-days" of work (See Walker, 1965, Chapter 1). This tremendous structural change, involving the reshuffling of tens of millions of households, was accomplished without the disruption that had attended Soviet collectivization. While it did not lead to any substantial increase in farm output, it undoubtedly facilitated government procurement of agricultural commodities.

India, on the other hand, eschewed radical changes in agricultural organization, limiting itself to some redistribution of land through purchases from large landowners. The first FYP, as already noted, placed considerable emphasis on agricultural investment in the form of irrigation, fertilizers, and improved seed. The results were not unimpressive—a 26 per cent increase in the output of food grains during the first FYP plan period, and 15 per cent during the second (Government of India, 1961, p. 36). Most of this gain was offset by increasing population, leading many Indians to advocate more far-reaching measures, including further land reform and the formation of producer cooperatives. Had the Chinese farm program proved successful, India might well have moved in the same direction, but, as things turned out, the Indians continued to follow a cautious policy. The third FYP provided for more irrigation, the extension of dry farming, and the expansion of voluntary development schemes to provide better contact with the individual farmer. The original draft of the fourth FYP assigned top priority to agriculture, in an effort to free the country from its dependence on food imports.

The Indian agricultural record has by no means been brilliant, but production has risen, on the average, $2\frac{1}{2}$ to 3 per cent a year since planning was initiated, without the violent fluctuations experienced by China (Lewis, 1964, pp. 151–52). India is still close to the margin of subsistence, and a bad crop year can lead to near-famine conditions, but the government is pinning its faith on the initiative of the individual peasant, in sharp distinction to the collectivism of China.

3. *Choice of technique.* It has been said of the USSR that "investment allocation within the nonfarm sector has often been economically irrational . . . nonfarm sector projects in favored branches must often have been inordinately capital intensive relative to those in less favored branches" (Bergson, 1964, pp. 269–71). This is no place to take up the thorny problem of the appropriate relationship between capital and labor in developing economies;[7] suffice it to say that if the Russians tipped the scales unwarrantedly in the direction of capital intensity, the Chinese, with relatively greater manpower resources and paucity of capital, tipped them still further.

There is no evidence that the Chinese were unhappy about this aspect of their investment program. It is likely that they had little alternative. The core of the first FYP, about half of all investment in heavy industry, consisted of 156 Russian-designed projects, many of them involving the movement to China of equipment for entire factories. Lacking engineers and designers, as well as skilled labor, it would have been difficult for the Chinese to reject the advice of the Soviet technicians sent in to help them construct and operate the new plants. The Russians might have sent in older equipment requiring a greater use of labor, or (if credits could have been arranged) such equipment might have been procured from other countries, but this would have required a degree of sophistication not commonly found in developing nations. Moreover, the nature of the new plants probably imposed considerable technological constraint; it is difficult, for example, deliberately to plan labor-intensive steel mills or machine assembly plants.

The Indians were not very different from the Chinese in this respect, having placed considerable emphasis on the development of modern steel and engineering industries. However, they were more prone than the Chinese to experiment with labor-intensive manufactures, particularly in textiles, where special equipment was designed for the purpose. That this policy has not been particularly successful is another story.[8]

7. For a good summary, see Meier, 1964, pp. 229–250.
8. Among the obstacles to realization of profitable labor-intensive manufactures have been the continuation of archaic methods of production and the necessity of maintaining quality standards for export goods. There has also apparently been considerable reluctance among Indian authorities to pursue this goal in a systematic way. See Lewis, 1964, pp. 58–62.

To conclude this section, it should be said that the Chinese industrialization drive of the period 1952–1957 must rank as one of the greatest concentrated efforts in world history, in terms of the number of people involved, the amount of material used, the capital put in place, and the increase in industrial output. The Communist leadership, emboldened by the radical transformation of their economy, sought to move to a new phase of development more consistent with their factor endowments. It is paradoxical that this new policy, as seemingly rational in the abstract as the old one was seemingly irrational, proved to be their undoing.

The Great Leap Forward

Industrialization via the capital-intensive route meant, among other things, that relatively few new jobs were opening up in the modern sector of industry. Employment in the modern sector (industry, water conservation, construction, transportation, and non-traditional trade) rose only from 30.2 million to 31.0 million from 1952 to 1957; in producer goods industries, the increase was from 2.7 million to 4.7 million (Emerson, 1965, pp. 128–43). Hordes of young workers were entering the labor market, making more urgent the provision of additional employment opportunities. The obvious answer seemed to be deployment of this great mass of under-employed manpower in work that would contribute to further development.

Indian planners faced the same pressure for action. Despite the industrial progress achieved during their first FYP, the number of unemployed actually increased from 1951 to 1956, and rural underemployment remained an unresolved problem. When the second FYP was adopted, it was estimated that 15.3 million new jobs would have to be found during the plan period to accommodate labor force growth and existing unemployment, entirely apart from making any inroads into rural underemployment (Government of India, 1956, pp. 110–12). The employment section of the plan ended on this pessimistic note: "in spite of concerted efforts for the mobilization of available resources and their optimum utilization as proposed in the second plan, the impact on the two-fold problem of unemployment and underemployment will not be as large as the situation demands" (*Ibid.*, p. 124).

Such caution was not characteristic of the Chinese, who at the time were in full flush with the apparent victory of their first FYP drive. Far from being willing to subscribe to the notion that their efforts would not be "as large as the situation demands," they sought to reach new and unprecedented heights. The essence of the new policy, termed the Great Leap Forward, was:

1. Industrial output was to rise by at least 25 per cent per annum. Chou En-lai defined a 20 per cent increase as a leap forward, a 25 per cent increase as a great leap forward, and a 30 per cent increase as an exceptionally great leap forward (Yeh, 1967, p. 350). By this definition, the new policy seemed moderate rather than extreme; who, after all, would be

satisfied with just a simple leap? That this implied a substantial increase over what had been claimed for the extraordinary 1952–1957 period was another matter.

2. The heavy industrial sector was to continue its rapid growth, even more rapidly than during the first FYP. Wherever possible, however, labor-intensive processes were to be employed instead of more capital. As a leading official put it:

> We cannot demand that all factories, mines and enterprises are equipped with modernized equipment. We must first consider before plunging into modernization and automation. At present we are doing just the opposite; some factories which need not be modernized are being modernized and thus much waste takes place (Hughes and Luard, 1959, p. 64).

3. New employment was to be provided primarily by small-scale shops using extremely labor-intensive techniques rather than by the large enterprises of the modern sector, and by construction projects in which human labor constituted the bulk of the input. At first it seems to have been the intent to expand mainly the producer goods industries, but it soon proved necessary to involve consumer goods production as well, since this sector was usually better suited to the desired techniques. This dual economy, in which labor-intensive and capital-intensive sectors would operate side by side, was dubbed "walking on two legs."

In fact, a dual economy is found in most underdeveloped countries, so that there was nothing basically new in the idea. What was novel was the scale on which the labor-intensive sector was to be expanded; this was not only encouragement to a traditional sector, but the creation of a whole new one.

4. Industrial management was to be decentralized in order to allow the fullest possible scope to local initiative. Up to 1957, all enterprises of any substantial size were under the direct control of the central government. During 1958, 80 per cent of the enterprises, including most of those producing consumer goods, were transferred to the supervision of local authorities. The number of directive targets, through which enterprise management was controlled, was reduced. However, the central government retained supervision over the large heavy industrial plants.

A few words on the basic economic rationale of the strategy are in order. That China was faced with an underemployment problem of serious dimensions can hardly be doubted. But the concept of underemployment is an exceedingly tricky one. The International Labour Office, which has devoted considerable study to the subject, concluded that "there is no general agreement as to how to go about measuring or even defining underemployment" (1964a, p. 25). Attempts at measurement have yielded widely varying results. The Food and Agricultural Organization, for example, has estimated that between 28 and 64 per cent of the agricultural workers of the Middle East, North Africa, and Southern Europe are surplus (*Ibid.*, pp. 28–29). India has paid particular attention to the problem, and its third FYP contains the following statement:

It is not easy to measure unemployment in an underdeveloped country. There is a tendency, especially among the self-employed, to share work between members of the family or the group. When the available work opportunities are spread too thinly even to provide tolerable means of livelihood, a part of the population migrates in search of paid employment. It is in relation to this section of the population that the term 'unemployed' can be used with some exactness. For the rest, one can only speak of under-employment for varying periods . . . At the present stage of development, it is difficult to determine the volume of underemployment with reference to 'norms' of hours per day to be worked by individuals or other similar criteria. It is more meaningful to judge the amount of under-employment by the extent of additional work an individual may be willing to take up (Government of India, 1961, p. 155).

On this basis, Indian unemployment in 1961 was estimated at nine million, and underemployment at 15 to 18 million. However, an even larger number of persons were occupied with "trivial, unproductive tasks which with existing methods keep them reasonably fully occupied but which, with little or no additional capital investment, could be reorganized so as to release them for other work without any reduction in output" (ILO, 1967a, p. 36). This is really the hub of the matter, for few people sit around in absolute idleness. On the contrary, many people whose marginal productivity must be close to zero work extremely hard; the trouble is that their work is economically meaningless.

The magnitude of Chinese unemployment in 1957 is just as questionable as the data cited above for 1952. Liu and Yeh put the figure at 32 million non-agricultural male workers, while Hou cites figures of from 9.6 to 18.3 millions (Liu and Yeh, 1965, p. 105 and Hou, 1968, p. 369). There are obviously basic conceptual differences between these estimates, but both record an increase between 1952 and 1957. No one has even attempted to estimate Chinese underemployment. If the Indian figure for 1961 were simply scaled up to reflect the larger Chinese labor force, a total of 29 to 35 million Chinese underemployed would be obtained.

As vague and imprecise as these figures are, they do provide some idea of the magnitude of the Chinese employment problem. It is not unlikely that, in some real sense, a portion of the labor force equivalent in size to the entire labor force of the United States was either producing nothing or working at a level of efficiency low even for China. It was tempting for the authorities to assume that if these idle hands were put to work, no matter how lowly the task, a positive net product must result.

The Chinese did not have to overcome the important institutional barrier present in democratic societies, namely, the presence of trade unions and other pressure groups. These groups make it possible that wages paid to workers on the kinds of projects envisioned by the Chinese will exceed their contribution to output. Since surplus labor had in any event to be maintained at a subsistence minimum, the principal additional cost would be the extra food required to make a productive worker out of an indolent man, plus the absolute minimum equipment necessary to enable him to carry out simple tasks.[9]

Once at work, the former unemployed would be expected to produce a surplus beyond their consumption requirements. This might be in the form of additional food (to the extent that capital construction in agriculture was involved), in consumer goods, in electric power and roads, or even in producer goods. A portion of the increased output could be used for industrial investment, so that there would be a simultaneous increase in consumption and investment. The small workshops would have low ratios of capital to output, which would mean that capital would not have to be tied up for long periods before the returns began to come in. Moreover, it was anticipated that much of the additional capital needed to put the scheme into operation would come from unexploited local resources, and would not affect the regular supplies already being mobilized by the government.

The draft of the second Chinese FYP, which disappeared without a trace amidst the excitement of the Great Leap Forward, anticipated increased industrial employment of only six to seven million for the entire period, although the labor force was increasing by that amount each year. The purpose of the Great Leap was not only to provide jobs for the new entrants into the labor market, but also to reduce existing unemployment. A special feature of the scheme was the location of most of the new workshops in rural areas, where their operation could be dovetailed with the peak labor requirements of agriculture.

A number of conditions had to be fulfilled for the successful operation of a plan of this nature. Among them are the following:

1. The product turned out by the new labor recruits must be of economic value. Unusable goods of substandard quality, or poorly planned construction projects, do not add to the real national product, no matter how much labor is expended in their production.

2. The capital allocated for the new labor-intensive sector should not impinge upon the resources otherwise available for investment; and if it does, the productivity of the capital in its new use must exceed that which would have obtained had it been invested in the modern sector. Small-scale labor-intensive operations can easily result in waste rather than in stretching available capital resources. It is particularly important to prevent the diversion of agricultural materials from factories to rural workshops.

3. The product of the new labor must exceed the additional costs involved in its mobilization and deployment. This includes not only additional food,

---

9. A man might be able to subsist on 1600–1700 calories a day, but he would need at least 2,500 to put in an effective day of work. Leibenstein, 1957, p. 64. The Chinese rationing system in 1955–1957 made the following food differential allowances (nonmanual workers = 100): light physical labor, 113; heavy physical labor, 145; exceptionally heavy physical labor, 178. Liu and Yeh, 1965, p. 48. The required equipment might be hand tools in manufacturing and baskets, slings, shovels, and pickaxes in construction.

but housing, medical, and sanitary facilities that may be involved if people are moved from their homes, particularly if the work must be done in large groups.

4. In the absence of standard wage payments, there must be adequate incentive to work. Speeches and slogans may suffice for a short campaign, but not over the long pull.

The outcome of the Great Leap is discussed below. But it is clear that these conditions were far from satisfied. Much of the product turned out by the tens of millions of workers involved proved to be of little or no economic value.[10] The most notorious example was that of the backyard blast furnaces, which produced iron of substandard and often entirely unusable quality. Many hastily dug canals and reservoirs destroyed natural irrigation and ruined land by raising the underground water level. Disruption of raw material supplies impeded operation of the modern sector. And, entirely unforeseen, the excessive withdrawal of farm labor created a serious shortage of help and contributed to the decline of farm output (C. Y. Cheng, 1963, pp. 141–42). Wholesale movement of people and inadequate provision for their sustenance appears to have undermined morale.

There can be little doubt that the Great Leap experiment in rapid economic development was a failure. Far from attaining new production records, China found itself in deep economic crisis by the end of 1960. Urban unemployment rose sharply, forcing many workers to return to their villages. In an unprecedented act, the authorities permitted thousands of people in Kwangtung province to cross the border into Hong Kong in 1962 because they could not be fed. Much of the gain from a decade of hard work had been dissipated.

Development Policy since the Great Leap Forward

Since the Great Leap Forward, the Chinese have followed a much more cautious policy. Though details are lacking due to a blackout of economic and statistical information, the main outlines of the new policy are clear. It was heralded by a declaration of the Communist Party in 1961, which laid down the main lines in the following terms:

1. In view of the fact that agricultural output was adversely affected by serious natural disasters during the past two years, the entire country in 1961 must concentrate its efforts on strengthening the agricultural front, on carrying out thoroughly the guide line of agriculture as the foundation of the national economy, of having the whole party and all the people participate (either directly or indirectly) in agriculture and in the production of food grains. The assistance provided agriculture from all other walks of life must be strengthened in order to strive to the utmost for better results in agricultural production. . . .

10. Employment in the material production branches of the economy rose from 30.9 million in 1957 to 47.9 million in 1958. Emerson, 1965, p. 128. It was reported that 100 million farm workers were put on earth-moving projects in the spring of 1958. C. Y. Cheng, 1963, p. 138.

3. The light industries must energetically overcome the difficulty of raw material shortage due to natural disaster, explore new sources of raw material, increase production, and do their best in guaranteeing the supply of necessities of living.

4. As to the heavy industries, the scope of capital construction should be appropriately reduced; the speed of development readjusted; and, based on previously achieved advances, a guideline of consolidation, reinforcement, and improvement is to be adopted. Strenuous efforts must be made to improve the quality of products, to increase their variety, to strengthen the weak links in the production system, to continue the development of technical transformation among the people, to economize on raw material, to lower the cost of production, and to raise labor productivity. (*People's Daily,* Jan. 21, 1961, p. 1).

Ever since, the emphasis has been upon agriculture. New industrial investment is justified only if it contributes to agricultural growth, as, for example, through the production of chemical fertilizers and farm machinery. Consumer goods industries, particularly those turning out goods for peasant consumption, receive priority over heavy industry. Rather than building new plants, the technology of old ones is to be modernized. In a sense, the Chinese seem to have embraced, at least for the time being, what we termed at the outset the strategy of balanced growth, except that the planners, rather than the market, are expected to guide the economy into activities which maximize current output.

Interestingly enough, the Indians were moving in the same direction at about the same time. The second Indian FYP had stressed the growth of the steel, cement, coal, machinery, and heavy chemical industries, at the expense of agriculture and light industry. Serious agricultural problems arose, and a balance of payments crisis necessitated postponement of a portion of the planned industrial investment. The third FYP, which was adopted in 1961, increased the investment share of agriculture. Mindful, perhaps, of the Chinese experience, the Indian planners proposed no crash program to absorb their 25 plus millions of unemployed and underemployed. They did indicate, however, how they hoped to alleviate the plight of this group by the following measures:

1. A program of rural works, involving irrigation, forestation, soil conservation, road development, and land reclamation, was to provide employment for 100,000 persons during the first year of the plan, with the total rising to 2.5 million by the last year. Pilot studies were to be undertaken to test the feasibility of such schemes, before large scale operations were attempted.

2. Labor-intensive techniques were to be given more attention in construction projects. This, it was hoped, would provide 2.3 million new jobs.

3. Although no wholesale expansion of village and small-scale industry was contemplated, this sector was to be strengthened by raising its productivity through training of employees, improved tools and equipment, government-provided technical advice, improved credit facilities, and better

organization of supply and marketing. This, it was hoped, would provide fuller employment for eight million people and full-time employment for about a million of the unemployed.

A critic of the employment aspects of the Indian five year plans wrote:

> Even when they are all put together, these indications of policy intent do not, to be sure, yet add up to an adequate or decisive idle-manpower mobilization effort. The plainest indication that they do not is that the Plan as it stands makes very little financial provision for their implementation . . . It represents an unfinished revolution in Indian development strategy that by mid-1961 had gathered enough momentum to permit the statement of some lines of policy that are decidedly novel by Second Plan standards. However, by the publication of the Third Plan, the proponents of the new approach to idle-manpower mobilization had not yet had time adequately to articulate, and to muster consent for, the financial and organizational corollaries that the approach logically implies. (Lewis, 1964, pp. 93–94).

The experience of China, however, may well justify the Indians in their caution. The dangers attendant upon hastily and poorly-conceived schemes were all too evident. There is no easy, rapid way to transform unskilled labor power into productive capital. This does not mean that there are no possibilities in this approach. But the lesson of the Great Leap Forward is that doing nothing may be less damaging than rushing in without careful preparation.

CHAPTER THREE

Development of the Industrial Sector

During the first decade of their power, particularly the years 1952 to 1959, the Chinese Communists concentrated almost single-mindedly on the construction of an industrial base. The underlying rationale of this policy was discussed in the preceding chapter. We are concerned here with the details of the execution of the policy.

We must be clear, however, that although priority was given to industrial development, the industrial sector did not loom large in relation to the total economy. If we define it to include factory production, mining, utilities, construction, and modern transportation and communications, its contribution to the domestic product in 1957, after the completion of the First FYP, was 28 per cent.[1]

At least until the demise of the Great Leap Forward, China was a heavy investor, compared with India and even with Russia at a comparable state of Soviet power. (See Table II-2 above.) The bulk of the investment went into fixed capital formation, particularly of an industrial character. It is estimated that 36 per cent of gross fixed investment in 1952 was allocated to industry, with this share rising to 52 per cent by 1957. (Yeh, 1968, p. 521) By comparison, 41 per cent of Soviet investment went into industry during its first FYP, and 43 per cent during the years 1952 to 1959 (Kaplan, 1953, p. 52 and 1963, p. 117). India put 25 per cent of total investment into industry during its first FYP (Malenbaum, 1959, p. 300). The Chinese concentration

---

1. This percentage is based upon the use of 1952 prices. If 1933 prices, which were not affected by the overvaluation of producer goods, are used, the ratio drops to 18 per cent. Liu and Yeh, 1965, p. 66.

on industrial investment must be considered all the more remarkable when the country's lack of overhead capital and housing are taken into account.

The lion's share of Chinese industrial investment, moreover, was allocated to producer goods. During the first FYP, the proportion varied from a low of 76 per cent in 1952 to a high of 87.7 per cent in 1955, leaving only a minor share for textiles, shoes, food processing, clothing, and other branches catering to the consumer (Chao, 1968, p. 577).

The Great Leap Forward saw a sharp increase not only in total investment, but probably in the share of industry as well. The aftermath of the Great Leap, however, brought about a sharp retrenchment, with heavy industry being particularly affected (Liu, 1968, p. 166). The tide appears to turn once again in 1963, when investment activity was stepped up, with particular emphasis this time on petroleum, atomic energy, chemical fertilizers, and synthetic fibers. Contracts were signed in 1964 for the import of six complete plants: four for the production of chemical fertilizers, one for alcohol, and an oil refinery, all from Western Europe (Lewin, 1964, p. 57). There was also a well publicized purchase of steel fabricating equipment from West Germany in 1966. These isolated incidents tell us nothing of the general level of industrial investment, but they testify to Chinese willingness to commit resources to the import of capital goods at a time when foreign exchange was being husbanded to help feed the population. It is very doubtful, however, whether the share of investment allocated to industry has approached in magnitude the levels achieved in the forced march to industrialization during the first decade of Communist rule.

The Soviet Assistance Program

Few facets of Chinese economic development have been the subject of more speculation than the economic relations between China and the Soviet Union. Some of the basic facts concerning the pattern of trade and aid between the two great Communist powers are still obscure, and there is controversy about the interpretation of others. There is one outstanding fact which is perfectly clear, however: without Soviet Russia as a source of supply from 1950 to 1960, China could not have made such great industrial progress. Whether the basic policy was wise is another matter; but, given the decision to industrialize, Soviet willingness to provide the sinews of heavy industry was crucial.

The Russians agreed initially to aid in the construction and reconstruction of 156 major industrial enterprises, including seven iron and steel plants, 24 electric power stations, and 63 machinery plants. They provided a broad range of services in connection with these projects: "The 156 projects under Soviet aid ranged from prospecting and surveying geological conditions, selection of factory sites, collecting data for the construction of factories, designing, supplying equipment, advising in actual construction, installation and operations, training personnel, supplying technical data and blueprints, down to manufacturing of new products."[2] The number of projects was expanded to 211, but amalgamation brought the total back down to 166 by the end of 1957. In 1958 and 1959 an additional 125 projects, scheduled for completion in 1964, were added, bringing the grand total up to 291.

How many of the projects were actually completed either before or after the deterioration of Sino-Soviet relations is not known. By the end of 1957, some 68 were finished, plus 45 more in 1958, and 41 in 1959 and 1960, or 154 in all (Li, 1964, p. 31). A Soviet statement in 1963 announced that "with the active assistance of the Soviet Union, People's China has built 198 enterprises, shops, and other projects equipped with up to date machinery." (*Pravda*, July 14, 1963). The Russians also asserted that "in April, 1965, the C.P.R. Government officially renounced cooperation with the USSR in constructing a number of industrial projects stipulated in the Chinese-Soviet 1961 agreement," (New York *Times*, March 24, 1966, p. 14). The various figures are not unambiguous, however, and a Western estimate puts the number of completed projects in 1960 at 130 (C. Y. Cheng, 1964, p. 47). Some of the others may have been completed subsequently without Soviet aid, but many were probably abandoned.

The completed Soviet aid projects were the very core of the Chinese industrialization program. On an aggregate basis, the original 156 projects called for a combined Soviet and Chinese expenditure of 11 billion *yuan*, which, together with an additional 1.8 billion *yuan* for the construction of ancillary projects, constituted 48 per cent of the total industrial investment planned for the First FYP (Chao, 1968, p. 570).

The importance of the Soviet program is borne out by a closer look at some of its principal components. The three major Chinese iron and steel complexes, at Anshan, Wuhan, and Pao t'ou, were equipped largely with Soviet machinery. The Anshan steel plant in Manchuria, the largest in the country, which had been built up by the Japanese and then stripped by the Russians at the end of the war, was rebuilt with modern Soviet equipment both in basic steel and rolling. The Wuhan mills in central China and the Pao t'ou mills in Inner Mongolia were built from the ground up to Soviet specifications (C. Y. Cheng, 1964, pp. 32–33 and Hsia, 1964, pp. 126–31). According to the Russians, Soviet project plants were producing 30 per cent of China's pig iron, 39 per cent of the steel, and 51 per cent of the rolled products in 1960 (Wu, 1965, p. 187).

The machinery industry was another in which Soviet aid counted heavily. A number of major enterprises, including the Changchun automobile plant, the tractor and ball bearing plants in Lo-yang and Harbin, and the Peking electronic tube plant, were outfitted with Soviet equipment. To reduce the period of gestation for these and other enterprises, entire plants rather than mere individual pieces of equipment were installed. About 60 per cent of the $3 billion of equipment and machinery imports from Russia during the years

2. Li Fu-chun, Chairman of the State Planning Commission, quoted by C. Y. Cheng, 1964, p. 29.

1950 to 1962 took this form. Its significance has been summarized by Eckstein as follows:

... the deliveries of complete plant installations represent a most important form of development assistance, using the latter term in a very special sense. It is development assistance in the sense that the same expenditure of foreign exchange on capital goods imports in its traditional form will almost certainly yield lower rates of industrial growth in the underdeveloped country which receives it than an equivalent outlay on complete plant installations. (1964, p. 145).

To cite a few more examples, the Lanchou oil refinery, with an annual capacity of about a million tons, which went into operation in 1959, was planned in the Soviet Union, and 85 per cent of its equipment was of Soviet origin. Of the oil industry as a whole it has been remarked that "projects for building several giant refineries, from designing to actual construction, have been mostly if not exclusively conducted with Russian and in some instances Rumanian assistance" (Chang, 1963, p. 29). Two major Chinese hydroelectric power plants and 19 of 41 thermal plants built by the end of 1957 were largely Soviet equipped.[3] The same is true of enterprises in the nonferrous metals, chemical, and coal industries.

The Soviet Union also provided China with technical assistance. About 11,000 Russian specialists worked in China during the decade ending in 1960, while 28,000 Chinese technicians and skilled workers went to the Soviet Union for training. Key personnel in new plants were particularly chosen for such apprenticeships; for example, 500 employees of the Changchun automobile plant were trained in the Likhachev auto works in Moscow (C. Y. Cheng, 1964, p. 41). China received 10,000 sets of specifications from Russia, ranging from machine designs to blueprints for large construction projects.

The importance of Soviet assistance began to decline as early as 1958. For the next two years, "except for a few ultra-modern projects for which Soviet aid had to be as thoroughgoing as before, the Chinese undertook the surveying and designing by themselves, relying on the Soviet Union for principal equipment instead of complete sets and for the supply of the most up-to-date design and product blueprints and other technical materials. The Soviet Union still had to send specialists to help in installation and first-stage operation." (Li, 1964, p. 31). By 1959, Chinese engineers were able independently to design and build a modern blast furnace, and the long Yellow River bridge at Chengchow was said to be entirely of Chinese design and construction.

Nevertheless, the break between China and the Soviet Union, which was marked by the recall of between 1,000 and 1,500 Soviet specialists in May, 1960, was a severe blow to Chinese industrial development. Projects under

---

3. The extent of Soviet assistance was acknowledged in the following terms: "Of the 1,421,000 kilowatts of electric generating equipment newly installed in the system of the Ministry of Electric Power from 1953 to 1956, only 112,000 kilowatts were made by China itself." Department of Industrial Statistics, 1958a, p. 125.

construction were halted, others were cancelled. Spare parts for Russian machinery became difficult to obtain. *People's Daily* summarizes the impact of the rift in the following terms:

In this period, we also encountered an unexpected difficulty. This was caused by the Soviet authorities, who in July, 1960, seized the opportunity to bring pressure to bear upon us and extended the ideological differences between the Chinese and Soviet parties to the sphere of state relations. They suddenly and unilaterally decided to withdraw all their experts, totalling 1,390, who were assisting China in its work, tore up 343 contracts and supplementary provisions concerning the experts, and annulled 257 items of scientific and technical cooperation. After that, they heavily slashed the supply of whole sets of equipment and crucial parts of installations. This caused heavy losses to China's construction work and dislocated its original plan for the development of the national economy, greatly aggravating our difficulty.[4]

A 1965 despatch from Peking, summarizing the contemporary state of Chinese industry, concluded that "allowance must still be made for the chaos caused by the Russians in many factories, the 1960 pullout was a staggering blow" (*New York Times*, Oct. 23, 1965, p. 45). A visitor to China in 1964, reporting his impressions of the industrial scene, wrote that "the outstanding one is the Russian sabotage in the summer of 1960, when Mr. Khrushchev withdrew the technicians in breach of their contracts and ceased all technical aid to China. It was only at the biggest blast furnace at Anshan that I was told frankly that the vacuum left by the Russians had not, in that case, been filled" (Wilson, Aug. 13, 1964, p. 272).

Economic relations between China and the Soviet Union did not terminate in 1960. The volume of trade between the two countries in 1965 was less than one-quarter of the 1959 level, however. Machinery imports declined dramatically from almost $600 million in 1959 to $27 million in 19f12, before rising a bit to $77 million in 1965. By 1965, more than two-thirds of China's trade was with non-Communist countries, though China continues to deal with the Soviet Union and other countries of Eastern Europe despite the deterioration of political relationships.

The Soviet Union did not provide China with free assistance or with large economic loans. The manner in which the assistance program was financed by China is considered below. However, without the willingness of the Soviet Union to sell machinery and equipment to China on a large scale, the rapid Chinese industrial growth of the decade 1950–1960 would not have been

---

4. *People's Daily,* December 4, 1963, quoted in C. Y. Cheng, 1964, p. 95. The following account was given of the effect of the breach on the Shih-Ching-Shan steel works: "The year 1960 was one in which both domestic and foreign demons created massive troubles. Geared to the needs of U.S. imperialism, the (Soviet) revisionists launched a sudden attack on China. At that time, the new steel mill had procured two sets of equipment from the Soviet Union and their installation was not yet completed. They perfidiously broke the contract, withdrew their experts, took away the main blueprints for installation, and stopped the shipping of undelivered machinery. This caused great hardship to the company which was at a loss regarding how to install the equipment and how to make the undelivered machinery." JPRS, No. 7, 1966, p. 12.

possible. For political as well as economic reasons, there was no alternative source of supply on anything like the scale required. China has since had cause to regret its industrial dependence upon the Soviet Union, but the paradox is that without agreeing to acceptance of such a status initially, industrialization would have been slow in coming.

The Growth of Industrial Output

Statistical measurement of industrial output is fraught with difficulties under the best of circumstances. When a country is undergoing great changes in economic organization and industrial structure, the problem of measurement becomes even more complicated (See Bergson, 1961, Introduction and Chapters 1–3). And when the handicap of inadequate and inaccurate data is added, attempts at such analysis may well appear either heroic or quixotic to an outside observer, depending upon his temperament. It should not be surprising, therefore, that those who have approached the task of measuring China's industrial progress should be cautious about the validity of their results, nor that the results of their investigations do not always coincide.

The biases to which the officially published Chinese production indexes are subject have been well summarized by Kang Chao (1965, Chapter 2). Some of them will be familiar to those who have followed the earlier discussion of Soviet production measurement; others are peculiar to China. Among them are the following:

a) The use of gross value of output as the standard of measurement is likely to result in increased double counting as the complexity of a nation's manufacturing processes grows.

b) The prices used by the Chinese in calculating gross value of output seem to impart a bias to the production index on several counts: (1) the 1952 constant prices which were used in estimating pre-1957 values overstated the real prices of producers' goods, the output of which grew most rapidly during the ensuing decade; (2) new products were put into the index at high starting-up prices and remained at the same level; (3) local industries often set their prices at relatively high levels compared with centrally controlled factories.

c) Definition and coverage of industries changed a great deal without adequate correction in the index.

The official index of Chinese industrial production for the period 1949 to 1959, together with several Western estimates, are shown in Table III-1. The indexes cover mining, manufacturing, timber production, and electric power output, and there are separate estimates for modern factory production and total output, including that of individual and cooperative handicraft workshops. For the period up to 1957, the Chao and Field indexes suggest a substantial overstatement in the official index, while the Liu-Yeh index is close to the official results. The Western estimates differ partly because of varying assumptions about the growth of individual industries (petroleum,

TABLE III-1
INDEXES OF CHINESE INDUSTRIAL OUTPUT, 1949–1959
(1952 = 100)

| | OFFICAL INDEXES | | CHAO INDEXES | | FIELD INDEXES | | LIU-YEH INDEX |
	Modern industry	Modern industry and handicrafts	Modern industry	Modern industry and handicrafts	Modern industry	Modern industry and handicrafts	Modern industry
1949	39.9	40.8	44.3	50.0	41.1	48.5	—
1950	52.0	55.7	57.7	63.3	54.6	61.1	—
1951	74.8	76.8	77.0	81.2	77.6	81.3	—
1952	100.0	100.0	100.0	100.0	100.0	100.0	100.0
1953	131.7	130.2	124.7	122.1	122.8	125.1	122.9
1954	153.6	151.4	141.6	139.4	143.1	143.0	142.2
1955	165.6	159.8	146.9	149.7	148.4	143.9	159.0
1956	217.1	205.0	182.2	179.4	188.3	178.3	210.8
1957	240.6	228.3	195.9	189.6	209.0	196.8	238.6
1958	n.a.	379.4	272.6	251.5	282.1	256.3	289.2
1959	n.a.	528.6	371.4	330.9	362.3	323.7	373.5

Sources:
Chao, 1965, pp. 88 and 89.
Field, 1967, p. 273.
Liu and Yeh, 1965, p. 66.
The underlying data for the official index can be found in Chen, 1967, Table 4-37.

paper, metal processing, daily use commodities) for which data were scanty and estimates had to be made on fragmentary evidence. (For an analysis of the reasons for these differences, see Field, 1967, pp. 279 *ff.*) The official index soars into the stratosphere during the Great Leap Forward, while the Western estimates end up in fairly close agreement. Because of the turmoil surrounding the Great Leap, all the estimates for the years 1958 and 1959, official and Western alike, must be taken with more than the customary grain of salt. (For an analysis of statistical difficulties during the Great Leap, see Li, 1962, Chapter 8).

Even allowing for deflation of the official claims, the rate of growth of industrial output for the decade is impressive. From 1949 to 1953, additional product was obtained largely through the restoration of war-damaged facilities. During the period 1953 to 1959, when new facilities were coming into operation, the Western estimates suggest a three-fold increase in the output of the modern industrial sector. How rapid a rate of expansion this was can better be appreciated by comparison with the achievements of India and the Soviet Union during the first seven years of their industrialization programs. The relevant data appear in Table III-2.

TABLE III–2

INDEXES OF INDUSTRIAL PRODUCTION FOR CHINA, INDIA, AND THE SOVIET UNION, FOR THE FIRST SEVEN YEARS OF THEIR RESPECTIVE PLANNING PERIODS

Year	China (1952=100)	China (1953=100)	India (1950–51=100)	Soviet Union (1927–28=100)
0	100.0	100.0	100.0	100.0
1	124.7	113.6	112.2	120.0
2	141.6	117.8	113.4	138.6
3	146.9	146.1	117.3	163.6
4	182.2	157.1	129.1	172.1
5	195.9	218.6	138.6	191.8
6	272.6	298.0	151.0	228.6
7	371.4	—	152.5	294.5

Sources:
China—Table III-1. The Chao index is used for the purpose of comparison.
India—Government of India, 1961, p. 733.
Soviet Union—Hodgman, 1953, p. 232. The Hodgman index shows a lower rate of growth for this period than the official Soviet index, but a higher rate than some alternative Western estimates.

In making comparisons of this kind, it is important that the base years should be suitable. Was the Chinese industrial system as fully restored to production by our year 0 as that of India or the Soviet Union? If not, a portion of the Chinese growth could be attributed simply to the restoration of existing facilities.

Field's observations on this point are pertinent: ". . . the large increase in output achieved in 1953 was a continuation of the rapid growth achieved during the period of economic rehabilitation and tends to suggest that the pre-Communist peak level of production was not reached until 1953. Because 1953 was really part of the period of economic rehabilitation, the average annual rate of growth of 12 per cent achieved during the years 1954–57 is a better measure of industrial growth in China than the rate for the First Five Year Plan as a whole" (1967, p. 275). For the entire plan period, industry grew at an average annual rate of 16 per cent.

Neither in the case of the Soviet Union nor that of India do similar recuperative factors appear to have been operating during the first years of their planning periods, so that the Chinese rate of growth should be discounted. This is done by shifting the base to 1953, yielding the alternative growth index shown in Table III-2.

There is no gainsaying the impressive performance of Chinese industry for the first decade of Communist power even if the 1953-based series is regarded as appropriate. In terms of the sheer magnitude of the effort, this was a remarkable achievement for a regime that inherited a backward, battered, war-weary economy, and had to improvise administrative machinery capable of coordinating the efforts of more than half a billion people. Whether the crash policy was wise in the long run, and whether the price paid by the Chinese people in terms of reduced consumption 'was too high, are questions that may legitimately be raised. But the fact remains that China emerged from the ranks of the underdeveloped nations and acquired a heavy industrial base in a very short period of time.

What might have happened had the Chinese foregone the Great Leap Forward and made their ideological peace with the Soviet Union can only be surmised. The rate of growth might have slowed in any event while the gains of the first decade were being digested. On the other hand, completion of the full catalogue of Soviet aid projects by the target year of 1967 would have expanded the industrial base greatly. Achievement of this target would have depended, however, on China's continued ability to invest heavily, which in turn hinged upon continued agricultural growth. Crop failures such as those which occurred from 1959 to 1961 would have curtailed investment and industrial growth in the absence of massive external economic assistance.

Unfortunately, what actually did occur after 1959 remains largely a matter for speculation because of the lack of data, official or otherwise, upon which estimates of output can be based. What may best be described as informed guesses yield the industrial production series shown in Table III-3. The index reflects the rapid decline that took place when the primitive steel furnaces, coal pits, cement plants, food processing, and repair workshops which burgeoned during the Great Leap Forward, were liquidated in the ensuing collapse.

Impressionistic observations tend to support the dismal picture portrayed by the aggregate index. A visitor to 15 industrial plants in 1964 reported that "much of the plant is working below capacity. Leaving aside the brand new chemical plant at Wuching, where production was apparently well up to capacity, only one factory would answer my question about plant utilization, and in that case a figure of 70 per cent was given to me.... It was fair... to conclude that the Anshan steel complex was working at about half its capacity and foreign observers in Peking who had recently visited the Wuhan steel works came to the same conclusion there" (Wilson, August 13, 1964, p. 272). By 1965, however, there had been a substantial recovery, and there was an official claim that the value of industrial output for 1966 was more than 20 per cent above that of 1965 (JPRS, No. 34, 1967). If true, this would have put 1966 output at the 1959 level, on the basis of Field's estimate in Table III-3.

TABLE III-3
CONJECTURAL ESTIMATE OF CHINESE
GROSS VALUE OF INDUSTRIAL
PRODUCTION, 1957–1965
(1959 = 100)

1957	60
1958	79
1959	100
1960	104
1961	69
1962	60
1963	66
1964	74
1965	81

Source:
Field, 1967, p. 273.

Expansion, however, was cut short by the Great Proletarian Cultural Revolution. It is too soon to assess the precise extent of the economic damage resulting from the frenzied events of 1966 and 1967, but it must have been substantial. There were widespread disorders in major industrial cities, including Shanghai, Anshan, Harbin, Lanchow, and Canton. Several blast furnaces were reported destroyed at Anshan. Civil war halted production in 2,400 factories in June and July, 1967, at Wuhan, the second largest steel center of the country (*The China Quarterly*, Oct.-Dec., 1967, p. 185). There were fixed battles at Taching, a major oil center (*Far Eastern Economic Review*, Sept. 28, 1967, p. 620). Trucks were commandeered by Red Guard units, leaving most transportation to coolie labor. Some ports were virtually shut down for a time. Railway service was continually interrupted by strikes, slowdowns, and armed clashes (*New York Times*, July 14, 1968, p. 6). The following account from the Chinese press itself indicates what happened at the Harbin Railway Bureau:

Before the power seizure, a small handful of intra-party power holders who followed the capitalist road deliberately created confusion... They practiced economism extensively.... Hoodwinked, a small number of workers quit production. Some went to Peking to 'petition'. Others quit their productive posts on the trains in the middle of journeys. At the time, there was nobody guaranteeing service in 15 locomotives, and some of these stopped for as long as 18 days. Railway transportation was once in a half-paralyzed state. A large number of weapons were damaged. Large quantities of goods were piled up in the goods compound. The speed of loading fell markedly... The command system was paralyzed. A succession of incidents took place in January (1967). Heavy damage was caused to the State. (U.S. Consulate General, Hong Kong, April 17, 1967, p. 12).

It may be years before the true magnitude of the post-Leap economic depression and the Cultural Revolution decline become known outside China. A major problem to be solved before the decline can be estimated is determining the height to which industry soared in the frenetic years of 1959 and 1960; on this there is sharp disagreement among Western economists (See, for example, Table III–1 above). It is not inconceivable that the Chinese themselves have only an imperfect knowledge of the details of the 1958–1962 period because of the strains it placed on their statistical system and because of what appeared to be administrative disorganization.

Let us assume that, by 1965, Chinese industry had attained the level of output indicated in Table III–3, and, further, that the advances of 1966 were cancelled out by the 1967 Cultural Revolution. This would mean an index of industrial production of 282 (1952 = 100) at the end of a 15 year period, based on Field's indices. By way of contrast, the industrial production index for India stood at 256 in 1966 (1951 = 100), also after a 15 year period (*Statistical Abstract of the Indian Union*, 1965, pp. 152–153). The Chinese lead was secured by the very rapid increase in output from 1957 to 1959. Indian growth, on the other hand, has been less spectacular but steady. Considering the tremendous cost of boom and bust to the Chinese people, it is not at all clear that the Chinese performance was superior to the Indian. An economist who visited the two countries in 1964 reported that "India's filthy cities have the genuine ugly stir of the industrial revolution, while China's incredibly clean cities are quiet, as if they wait for someone's orders. China is probably ahead agriculturally but India is in the lead industrially; all in all, it is India which gives the visitor more feeling of economic progress afoot" (*Far Eastern Economic Review*, April 8, 1965).

The Pattern of Industrial Expansion

Statistics of Chinese output for a number of industrial products for the years 1952 to 1957 are shown in Table III–4. For a few commodities, the percentage of increase is very high because they were virtually unproduced in 1952. Producer goods tended to show more rapid rates of increase than consumer goods. Woolens and paper constituted apparent exceptions to the rule, but much of the paper may have been produced for industrial rather than consumer uses.

Some light may be thrown on the nature of the precise industrial expansion path chosen by the Chinese by comparing it with that adopted by India. As has already been pointed out, the overall rate of industrial growth in China during the nineteen-fifties was considerably higher than that of India. But it is possible to compare the relative growth of various industries for periods in which total output increased by roughly similar amounts. The Indian index of industrial production was 194 in 1960–1961 (1950–1951 = 100), (Government of India, 1961, p. 64), which is close to the Chao estimate of 1959. (1952 = 100) for 1957 China. Table III–5 compares the percentage growth in output of as many industrial commodities as the data permit, covering these two periods. It enables us to see some of the major differences in industrial emphasis between the two countries.

The final column in Table III–5 shows the growth of output for a number of Chinese products, using the increase in Indian output for the same commodities as a base for the periods specified. Despite the fact that industrial output as a whole rose by about the same amounts in both countries, the Chinese steel industry expanded much more rapidly than that of India. This is also true of a number of other heavy industrial products. On the other hand, the Indian consumer goods industries grew more rapidly than the Chinese, relative to total output. One interesting fact is that by this yardstick the Chinese chemical industry, and chemical fertilizers in particular, suffered from neglect during the first FYP.

The data underline the bias of the first Chinese FYP toward heavy industry. It must be remembered that India, too, was making strenuous efforts to establish a heavy industrial base, and achieved substantial increases in the output of capital goods. Steel output more than doubled and electric power capacity rose by 150 per cent during the first planning decade. But to the Chinese planners, the steel, electric power, coal, and machinery industries had the highest priority, since they were considered absolute requisites for economic development.

Official Chinese output data for a reduced list of products for the period 1957–1959 are shown in Table III–6. Even if one were to concede the possibility of substantial increases in output during this period from newly commissioned factories begun during the first FYP, on the one hand, and from the many new workshops that were established during the Great Leap Forward, on the other, some of the figures appear fanciful. The Chinese government admitted that 30 per cent of the iron and steel produced in 1958 was not of commercial grade (Liu and Yeh, 1965, p. 116). The same must have been true of other commodities. It seems hardly likely that the production of cotton yarn, which is so dependent upon crop results, could have almost doubled in two years, while it only increased 29 per cent during the first FYP. The data in Table III–6 exemplify the problem of estimating the effects of the Great Leap Forward on Chinese industry.

TABLE III-4

INDUSTRIAL OUTPUT IN CHINA, BY PRODUCT, 1952–1957

Heavy industry and machinery	1952	1957	Per cent increase, 1952 to 1957
Pig iron (thousand tons)	1,900	5,936	212
Steel (thousand tons)	1,348	5,350	297
Electric power (million kw. hours)	7,260	19,340	166
Coal (thousand tons)	66,490	130,000	96
Crude petroleum (thousand tons)	436	1,458	234
Metal cutting machines (units)	13,734	28,000	104
Diesel engines (h.p.)	27,261	609,000	—[a]
Electric motors (thousand kw.)	639	1,455	128
Locomotives (units)	20	167	—[a]
Freight cars (units)	5,792	7,300	26
Merchant vessels (thousand dwt. tons)	16	54	238
Bicycles (thousand units)	80	806	—[a]
Cement (thousand tons)	2,860	6,860	140
Timber (thousand cu. m.)	11,200	27,870	149
Sulfuric acid (thousand tons)	190	632	233
Soda ash (thousand tons)	192	506	164
Caustic soda (thousand tons)	88	198	125
Chemical fertilizer (thousand tons)	181	631	249
Paper (thousand tons)	539	1,221	127
Tires (auto) (thousand units)	417	879	111

Light industry

Rubber footwear (thousand pairs)	61,690	128,850	109
Cotton yarn (thousand bales)	3,618	4,650	29
Cotton cloth (million metres)	3,829	5,050	31
Woolen fabrics (thousand metres)	4,233	18,170	329
Woolen yarn (tons)	1,980	6,133	210
Cigarettes (thousand crates)	2,650	4,456	68
Edible vegetable oils (thousand tons)	983	1,100	12
Sugar (thousand tons)	451	864	92
Salt (thousand tons)	4,945	8,277	67
Wheat flour (thousand tons)	2,990	4,220	41
Matches (thousand cases)	9,110	10,250	13
Aquatic products (thousand tons)	1,666	3,120	87

Note:
[a] Percentage increase not meaningful because of very low initial base.

Sources:
N. R. Chen, 1967, Table 4.6.
Liu and Yeh, 1965, pp. 454–455.

TABLE III-5
INCREASES IN OUTPUT OF SELECTED CHINESE AND INDIAN INDUSTRIAL PRODUCTS FOR SPECIFIED PERIODS
(in per cent)

Commodity	China (1952–1957)	India (1950–51 to 1960–61)	Chinese growth relative to Indian growth (Indian growth = 100)
Steel	297	120	248
Electric power	166	148	112
Coal	96	69	140
Electric motors	128	600	21
Bicycles	908	940	97
Cement	140	215	65
Sulfuric acid	233	267	87
Soda ash	164	222	74
Caustic soda	125	809	15
Chemical fertilizer	249	1122	22
Automobile tires	111	50[a]	222
Paper	127	207	61
Cotton cloth	31	38	82
Sugar	92	168	55

Note:
[a] 1955–1956 to 1960–1961.
Sources:
India: Government of India, 1961, pp. 77–81.
China: N. R. Chen, 1967, Table 4.6.

TABLE III-6
OUTPUT OF SELECTED INDUSTRIAL PRODUCTS IN CHINA, 1957 TO 1959

	1957	1958	1959
Pig iron (million tons)[a]	5.9	13.7	20.5
Steel (million tons)	5.4	8.0	13.4
Electric power (billion kw. hours)	19.3	27.5	41.5
Coal (million tons)	130	270	348
Crude petroleum (thousand tons)	1,458	2,264	3,700
Metal cutting machines (thousand units)	28	50	70
Cement (million tons)	6.9	9.3	12.3
Timber (million cu. metres)	27.9	35.0	41.2
Chemical fertilizer (thousand tons)	631	811	1,333
Paper (thousand tons)	1,221	1,630	1,700
Cotton yarn (thousand bales)	4,650	6,100	8,250
Cotton cloth (million metres)	5,050	5,700	7,500
Sugar (thousand tons)	864	900	1,130
Salt (thousand tons)	8,277	10,400	11,040

Note:
[a] Includes native iron. Of the total for 1958, 4.2 million tons were produced by indigenous methods
Source:
N. R. Chen, 1967, Table 4.6.

Some very rough estimates of post-Leap industrial production are given in Table III-7. The one industry that appears to have done at all well is chemical fertilizer, which was given the highest priority in the post-Leap period. Steel production in 1966 had just about regained the 1959 level, but the following statement from a Communist source may be indicative of what happened in 1967: "Owing to sabotage by the handful of party people in authority taking the capitalist road, the output of the Tsingtao Steel Mill dropped to the lowest point in its history just prior to the seizure of power. After takeover, the workers pushed the output of 15 products to an all-time high and fulfilled all the targets in the February plan of the mill" (JPRS, No. 184, 1967, p. 91). A group of Japanese experts who toured China in 1966 expressed the opinion that the productivity of the Chinese steel industry was only about 10 per cent that of the Japanese (JPRS, No. 29, 1967).

It is possible to argue about specific estimates,[5] but the main lines are reasonably clear. After the failure of the Great Leap, China de-emphasized heavy industry and stressed those branches of industry which could make the greatest contribution to raising agricultural output. There was some thought that the third FYP, scheduled to begin in 1966, would usher in renewed stress on heavy industry, but there is no evidence that this actually occurred.[6]

Handicraft Production

Reference has already been made to one of the major differences between the pattern of industrial development in China and the Soviet Union, namely, the treatment of handicraft industry. In the Soviet Union, this sector was systematically dismantled. In China, it has been maintained as an adjunct to factory production.

For statistical purposes, the Chinese divided the handicrafts into three groups: a) those conducted in conjunction with farming, including both subsistence production and the production of commodities for sale; b) individual and cooperative handicrafts carried on independently; and c) handicraft workshops. The second category, principally located in rural areas, produced mainly garments, cloth, thread, and small metal and wood objects.

5. Professor Ta-Chung Liu has indicated some scepticism with respect to the 1965 estimates: "Reservations seem desirable on these figures. Take the figures for steel, cement and timber as examples:

	1957	1965
Steel (million tons)	5.4	11
Cement (million tons)	6.9	9
Timber (million cu. metres)	27.9	36

Can these data be justified by any reasonable estimate of the relative magnitudes of investment in these years? I would rather doubt it. Yet these commodities are mainly for investment." Letter to the authors, April 30, 1968.

6. Some of the principal economic planners of China came under severe attack during 1967: Po I-Po, Chairman of the State Economic Commission; Li Hsien-nien, Minister of Finance; and Li Fu-chun, Chairman of the State Planning Commission. What effect this had upon economic planning is not known.

TABLE III-7
ESTIMATED OUTPUT OF SELECTED INDUSTRIAL PRODUCTS IN CHINA, 1960–1966

	1960	1961	1962	1963	1964	1965	1966
Coal (million tons)	325	180	180	190	200	210	(250)
Steel (million tons)	15.2	12.0	8.0	9.0	10.0	11.0	(12.2)
Chemical fertilizer (thousand tons)	2,480	1,450	2,120	3,000	3,600	4,600	
Crude petroleum (thousand tons)	4,500	4,500	5,300	5,900	7,000	8,000	
Cement (million tons)	13.5	6.0	6.0	7.0	8.0	9.0	(11.8)
Timber (million cu. metres)	33.0	27.0	29.0	32.0	34.0	36.0	
Paper (thousand tons)	2,130	1,000	1,000	1,000	1,500	1,500	
Cotton cloth (million metres)	6,000	3,000	3,000	3,300	3,600	3,900	(5,900)
Sugar (thousand tons)	920	700	480	540	1,100	1,500	

Sources:
1960–1965: Field, 1967, pp. 293–294.
1966: *New York Times*, May 23, 1967, attributed to Western analysts in Hong Kong.

Development of the Industrial Sector

The handicraft workshops were primarily in urban areas and differed from modern plants chiefly in size (Schran, 1964a, pp. 156–59).

The data in Table III–8 illustrate the continued importance of handicrafts under the first Chinese FYP. Although the level of primitive peasant production remained stable, the more advanced organizational forms almost doubled their output value from 1952 to 1957. The two latter categories contributed 36 per cent of gross value of industrial output in 1952 and 29

TABLE III–8
GROSS VALUE OF HANDICRAFT PRODUCTION IN CHINA, 1952–1957
(million *yuan* at 1952 prices)

	1952	1953	1954	1955	1956	1957
Agricultural sector						
Processing for own use	7.6	8.3	8.4	8.8	n.a.	n.a.
Work for others	2.3	2.4	2.5	3.0	n.a.	n.a.
Total	9.9	10.7	10.9	11.8		
Industrial sector						
Individual and cooperative	7.3	9.1	10.5	10.1	11.7	13.4
Workshops	5.0	6.8	7.5	7.7	8.3	9.4
Total	12.3	15.9	18.0	17.8	20.0	22.8

Source:
N. R. Chen, 1967, Tables 4.37 and 5.93.

per cent in 1957, (N.R. Chen, 1967, Table 4.37), a surprisingly consistent proportion when one recalls the rapid growth of factory production. In the Soviet Union, on the other hand, the share of small scale industry (which is roughly comparable to the non-farm Chinese handicrafts) fell very sharply during the first FYP.

India has tended to favor the handicraft industries to an even greater degree than China. The output of small enterprises rose by 38 per cent from 1950–51 to 1963–64, and, despite continuing emphasis on large scale industry, contributed 40 per cent of the total value of manufacturing output in the latter year (although the *relative* share of small scale industry, which was 62 per cent of the total in 1950–51, declined over the period. (*Statistical Abstract of the Indian Union*, 1965, p. 42). The rationale for the continued support of handicrafts, as set forth by the Indian planners, also must have been present in the minds of the Chinese to counteract the force of the Soviet model:

The objectives of the programmes for [village and small scale] industries ... are to create immediate and permanent employment on a large scale at relatively small capital cost, meet a substantial part of the increased demand for consumer goods and simple producers' goods, facilitate mobilization of resources of capital and skill which might otherwise remain inadequately utilized and bring about integration of the development of these industries with the rural economy on the one hand and large-scale industry on the other... With improvement in techniques and organization, these industries offer possibilities of growing into an efficient and progressive decentralized sector of the economy providing opportunities of work and income all over the country. (Government of India, 1961, p. 426).

The distribution of Chinese handicraft production by type of product is shown in Table III–9 for 1954, the only year for which detailed statistics are available. Among the important products not included in the list are milled rice, flour, tea, paper, and cotton cloth. The nature of the commodities produced in the handicraft trades illustrates how important a role they play in the satisfaction of consumer demand.

TABLE III–9
STRUCTURE OF CHINESE HANDICRAFT PRODUCTION, 1954
(Per cent of gross value of output)

	Per cent
Gross value of total handicraft production	*100.00*
Gross value of 13 important trades	48.75
Needle trades	13.62
Cotton spinning	8.19
Bamboo, rattan, coir, straw	6.17
Metal manufacturing	6.03
Wood processing	5.97
Edible oils and fats	3.09
Sugar refining	1.66
Leather manufacturing	1.23
Specialty handicrafts	0.93
Weaving	0.87
Pottery	0.57
Coal and charcoal mining	0.30
Silk reeling	0.12

Source:
Schran, 1964a, p. 168.

Official data record a sharp increase in handicraft production during the Great Leap Forward. Many of the hastily constructed workshops must have been closed down during the following years. However, with the de-emphasis of heavy industry, the handicrafts appear to have received some encouragement. Workshops were given greater freedom to buy raw materials and sell their final products on the open market (Stanford Research Institute, 1964, pp. 60–61). Currently, handicraft production may well constitute a larger proportion of total industrial output than in 1957. The smaller shops are nearer supply sources than large factories and in a better position to capitalize on temporarily favorable supply conditions.

The relative capital and labor endowments of China and India made the retention of the handicraft sector a matter of economic common sense. To the extent that additional output can be secured with small inputs of capital, there is a net gain to the economy, assuming that the labor involved otherwise would be unemployed. However, handicraft production must be organized in a manner that conserves capital and materials. This condition was not met during the Great Leap Forward.

Not every industry is suited to handicraft production. Where economies of scale are important, or capital equipment is indivisible, the factory usually has a clear advantage over the handicraft workshop. Considerations of quality and design of product may be important as well. A writer who has studied Asian cottage industries warns: "Once the difference in the patterns of production functions is taken into consideration, it will become evident that text-book type discussion on the problems of choice of techniques or scales of production in complete isolation from the choice of industries is unrealistic; the actual choice of the former is always influenced or, sometimes, even governed by the choice of the latter and, speaking more realistically, both choices are interdependent" (Ishikawa, 1966, p. 42). Part of the reason for the failure of the Great Leap Forward may have been the choice of the wrong industries for the newly established handicraft workshops. Heavy industry, which was favored during the Great Leap, is not generally suited to small scale production.

Whether handicrafts (apart from subsidiary farm production) will continue to play a major role in the Chinese economy is far from obvious. If a greater share of investment is allocated to the consumer goods industries, the traditional stronghold of small scale manufacture may crumble before the onslaught of the machine. Workshop production will survive only to the degree that its productivity can be raised by better organization of supplies and marketing, by injections of capital, by quality control, and by training of labor. India, which has made more strenuous efforts in this direction than China, has not had conspicuous success in making the handicraft sector more viable (See for example, Bhalla, 1956, p. 147 and 1964, p. 620).

The Choice of Technique in Manufacturing

Handicrafts aside, the Chinese planners had the option of industrial development through large, modern factories employing automatic machinery and relatively little labor, or through the application of large amounts of labor to less complicated and cheaper capital equipment. The ratio of available factors of production would seem to have dictated the latter alternative, but the former was in fact adopted in the first decade of the Communist regime.

To some extent, the Chinese had no choice. The Soviet aid projects came equipped with the latest Soviet machinery. The Russians might conceivably have sold old machinery requiring relatively more labor to China and replaced it with the new machinery which was being produced for Chinese account. However, this would have meant dismantling plants in operation, since there is no evidence of a surplus of capital equipment in the Soviet Union during the 1950's. Moreover, there are many pitfalls in the way of efficient use of second-hand machinery in underdeveloped countries: repair facilities must be adequate; spare parts must be available; training and operating manuals should ordinarily accompany the equipment; and most important, the machinery must be sufficiently modern and in good enough physical condition to compete with new equipment. We do not know what alternatives were explored in the negotiations between China and Russia, but there is no indication in the available information that anything but new equipment was ever considered by the Chinese.

Apart from the Soviet aid projects, China built a number of large enterprises on its own. A total of 2,056 major factory and mining enterprises (including the Soviet aid plants) were completed or under construction between 1953 and 1958, of which 1,037 were either wholly or partially in operation by the latter year.[7] Presumably the Chinese had a greater range of technological choice in the plants which they designed and built themselves.

There have been no industry studies in sufficient depth to indicate whether China took advantage of freedom of choice to go further along the route of labor-intensive processes. There was some questioning in 1957 of "gigantism," and it was suggested that smaller plants might have advantages over larger ones in speed of construction, lower capital-output ratios, and greater ease of domestic equipment production. There was no implication, however, that the small plants were to be outfitted with anything but machinery of the latest design (Chao, 1968, pp. 570–71). On the contrary, prior to 1958, new enterprises appear to have been constructed without much attention to employment potential, as the following indicates:

In the First Five Year Plan period, industrial employment in general grew more slowly than in the preceding period. The rate of increase in [gross value of] industrial output continued to be higher than that for employment . . . In the technologically advanced branches of industry, increased investment actually tended to reduce the rate of growth in employment, mainly because of technological displacement of labor (Emerson, 1965, p. 103).

The Great Leap Forward ushered in an abrupt about face in technological policy. Labor intensity in manufacturing became the watchword. However, lack of advance preparation and haste in construction of the 40,000 industrial projects completed in 1958 doomed the new approach from the start. The effect of the Great Leap on technological ratios is apparent from the data

7. N. R. Chen, 1967, Table 3.42. The term "major project" is not defined in the source, but from the context it appears to include the so-called "above-norm" projects, which are projects costing more than certain specified amounts varying among industries. For example, the norm was 3 million *yuan* for food plants and 10 million *yuan* in the iron and steel industry. *Ibid.*, p. 69.

TABLE III-10
SOME TECHNOLOGICAL RATIOS IN CHINESE INDUSTRY,
1952, 1957, AND 1958

	1952	1957	1958
Consumption of electricity per worker (kwh)	1430	2606	942
Capacity of power machinery per worker (hp)	1.73	3.01	1.25
Fixed assets per worker (*yuan*)	3525	5168	2026

Source:
Chao, 1968, p. 572.

in Table III-10. While the figures must be regarded as very approximate, they do give some idea of the tremendous capital dilution that took place in a single year. It is doubtful whether the modern enterprises altered their technological ratios, so that the new enterprises must have been very heavily weighted toward labor intensity.

Collapse of the Great Leap seems to have induced a return to the earlier policy. Uneconomic enterprises were closed, stress was placed upon raising quality of output, lowering of production costs, and economizing of materials. New investment has been concentrated in the production of chemical fertilizers and synthetic fibers, which do not lend themselves to labor-intensive production techniques.[8]

India, like China, has stressed technical modernity rather than labor absorption in expanding its modern industrial sector. The first Indian FYP stated that "the possibilities of increasing employment in large-scale manufacturing industries directly are limited, especially when the emphasis has to be on expanding producer-goods industries. For absorbing all, or a large proportion, of the increase in working population each year in non-agricultural occupations, reliance will have to be placed mainly on small scale and cottage industries involving comparatively small capital investment" (Government of India, 1953, p. 25). A decade later, the Third Five Year Plan, while noting the necessity of increasing employment opportunities, stated that "increase in investment and capacity does not lead to a proportionate growth of employment because new processes, specially in large-scale manufacture, have generally to be based on high productivity techniques... In certain branches of industry it is essential to adopt the scale and methods of production which will yield the largest economies" (Government of India, 1961, p. 157).

One would be hard put to distill from the Chinese experience any definitive rules for successful industrial development, or to identify dangers to be avoided. Like many other underdeveloped countries, the Chinese choice from 1952 to 1957 was limited by what it could get from outside in the way

---

8. It was reported in 1966 that in seeking to purchase industrial equipment from Japan, the Chinese were interested only in new equipment. JPRS, No. 29, 1967.

of equipment and technical advice.[9] There were neither the trained manpower nor the facilities needed for the development of equipment more suited to the relative availabilities of capital and labor. The second phase, the period of the Great Leap Forward, proved nothing except that the application of military mobilization techniques to industrial development is not likely to lead to success. In the years that have elapsed since 1959, the Chinese have had to retrench. Presumably they also have had time for more careful examination of industrial policy. No longer reliant on the Soviet Union, their choice of models has widened. A large-scale program for training engineering personnel has given them greater flexibility in independent design. It would not be surprising to discover that China has innovated in the direction of more labor-intensive manufacturing equipment. It is one of the few countries with sufficient incentive and resources to make this feasible. Nor is it beyond the realm of possibility that China will emerge eventually as a supplier of capital equipment to smaller countries in similar situations.

The Location of Industry

The industry of pre-Communist China was concentrated mainly in the coastal areas of the country. Foreign investment was placed in the treaty ports, and Chinese capital tended to cluster there as well, attracted by political stability and the existence of an economic infrastructure (Hou, 1965, pp. 155 and 217). When the first FYP was initiated in 1953, 77 per cent of the gross value of industrial output originated in the coastal areas.[10] The proportion was even higher for some major products: 82 per cent for cotton yarn, 88 per cent for cotton cloth, and 80 per cent for metal manufacturing (N. R. Chen, 1967, Tables 4.20 and 4.52).

This pattern of location was considered highly undesirable by the Chinese planners on a number of counts. First, there was the obvious matter of national security, the coastal areas being the most vulnerable to military attack. Factory location usually had little economic relationship either to the domestic sources of raw materials and fuel, or to the domestic markets for finished goods. The cotton textile and flour mills of Shanghai and Tientsin and the large oil refinery at Dairen operated mainly on the basis of imported raw materials, with a substantial part of their product being exported, while interior raw material-producing areas had little industrial productive capacity. Away from coastal areas, there was only one heavy industrial base, which the Japanese had developed in the Liaoning Province, and several additional centers of light industry.

9. "By February, 1960, the Soviet Union had sent to Communist China about 10,000 sets of specifications, including more than 1,250 designs for capital construction, about 4,000 blueprints for the manufacture of machinery and equipment, and more than 4,000 sets of technological and departmental specifications." C. Y. Cheng, 1965, p. 202.

10. N. R. Chen, 1967, Table 4.43. The coastal areas include the cities of Peking, Tientsin, and Shanghai, and the provinces of Hopei, Liaoning, Shantung, Kiangsu, Fukien, Kwangtung, and Chekiang.

The First Five Year Plan envisioned the spread of industry to the interior of the country. Two new steel enterprises were begun, one at Wuhan in Central China, the other at Pao-t'ou in Inner Mongolia. By 1961, the former was turning out 8 per cent of national pig iron production, but the latter was just beginning to operate.[11] Some 55 per cent of First FYP industrial investment was allocated to non-coastal areas, and since much of the coastal investment was for reconstruction of existing facilities, about three-quarters of all new plant investment was in the interior (Chao, 1968, p. 559). This policy was not without cost, for returns to investment would probably have been greater had new plants been concentrated in the developed areas where transportation facilities, utilities, and trained labor were already available.

While relocation opportunities were greatest for heavy industry because this sector was allocated the lion's share of new investment, the First FYP called for speedy establishment of new cotton textile plants in the raw materials producing areas in order to "change the present irrational distribution of China's textile industry." (*Major Aspects of the Chinese Economy Through 1956*, p. 261). Between 1952 and 1956, the proportion of cotton yarn manufactured in coastal areas fell from 82 to 72 per cent, and of cotton cloth, from 88 to 80 per cent (N. R. Chen, 1967, Table 4.20). The three major textile cities of Shanghai, Tientsin, and Tsingtao saw their share of the nation's spinning capacity decline from 56.7 per cent to 47.2 per cent between the same years, with an even more rapid decline in weaving, from 65.1 to 46.5 per cent (*Major Aspects of the Chinese Economy Through 1956*, p. 311). According to the Chinese themselves, however, at the end of the First FYP, "the imbalanced geographical distribution of the textile industry inherited from old China has not yet been basically eliminated" (*Ibid.*, p. 313).

The structure of manufacturing output for Shanghai and 13 provinces is shown in Table III–11 in as great detail as the availability of the data permits. Of the provinces shown, Liaoning had the greatest concentration of heavy industry, with not far from half of its total output in the steel and machinery industries. Eastern China was specialized in the textile and food processing industries. Light industry continued to predominate in the other major areas, though the establishment of the heavy industry base at Wuhan is reflected in the relative decline of light industry in Hupeh Province.

In general, up to the end of the First FYP period, shifts in the location of industry, in terms of output, were still moderate. When all the new facilities planned in the pre-Great Leap period are finally completed, the changes will be more notable. Hupeh and Inner Mongolia will begin to challenge the traditional dominance of Manchuria in heavy industry. The original list of Soviet aid projects included major machinery enterprises at Sian, Lanchow, Loyang, and Chi-chi-haerh, which were to become machinery producing

---

11. Hsia, 1964, pp. 130–131. In 1965, the Anshan works in Manchuria still accounted for at least half of national steel output. Wuhan had an estimated output of 1.5 million tons and Pao-t'ou of one million tons (pig iron, not steel). *Far Eastern Economic Review*, March 31, 1966, p. 623.

TABLE III–11
THE INDUSTRIAL STRUCTURE OF MANUFACTURING
INDUSTRY IN CHINA, BY PROVINCE, 1952 AND 1957
(Per cent of provincial totals, measured in gross
value of output)

	1952	1957
Manchuria		
Liaoning		
Iron and steel	n.a.	23.3
Machinery	n.a.	20.4
Textiles	n.a.	7.3
Kirin		
Machinery	2.3	16.0
Lumber	20.9	9.8
Paper	14.3	14.5
Food	25.0	16.9
Heilungkiang		
Machinery	6.2	11.6
Lumber	29.5	18.4
Food	24.5	18.5
Eastern China		
Shanghai		
Machinery	n.a.	18.0[a]
Textiles	n.a.	32.6[a]
Kiangsu		
Machinery	n.a.	13.9[a]
Textiles	n.a.	32.0[a]
Food	n.a.	19.6[a]
Anhwei		
Machinery	0.1[b]	5.7[a]
Chemicals	0.3[b]	9.5[a]
Textiles	12.8[b]	6.7[a]
Food	50.8[b]	29.8[a]
Chekiang		
Machinery	2.9	5.4
Textiles	30.8	27.6
Food	45.4	47.2
Central China		
Hupeh		
Iron and steel	2.4[b]	6.1
Machinery	2.2[b]	7.2
Textiles	27.3[b]	21.2
Food	38.0[b]	31.0
Hunan		
Non-ferrous metals	10.1[b]	9.6
Machinery	3.9[b]	10.3
Chemicals	0.6[b]	8.9
Textiles	17.1[b]	11.2
Food	39.6[b]	27.7
Kiangsi		
Non-ferrous metals	n.a.	16.5
Textiles	n.a.	17.0
Food	n.a.	30.9

TABLE III–II—CONTINUED

	1952	1957
Southern China		
Kwangtung		
Metal processing	6.5	11.4
Textiles	13.9	13.0
Food	52.3	44.5
Kwangsi		
Non-ferrous metals	42.4[b]	17.4
Metal processing	7.0[b]	9.4
Food	28.3[b]	44.6
South-Western China		
Szechuan		
Metal processing	7.8	12.2
Textiles	21.6	9.8
Food	33.5	39.3
Kweichow		
Non-ferrous metals	4.5	7.1
Metal processing	3.0	10.1
Textiles	25.3	10.1
Food	25.1	33.7

Notes:
[a] 1958
[b] 1950

Source:
N.R. Chen, 1967. Table 4.56.

centers. The Chinese Communists seem determined to push industry westward, and it is not unlikely that the new industrial areas will receive a preponderant share of investment in the future, and will challenge the long dominance of the coastal areas.

The Fuel and Mineral Industries

Coal is China's principal fuel by a wide margin. Even if one discounts the claimed production for 1958 and 1959, it is evident from Table III–12 that coal production expanded rapidly under the Communist regime.

The original list of 156 Soviet aid projects included 27 in coal mining. About 12 per cent of total planned investment during the First FYP was allocated to this industry, and 179 coal projects started between 1953 and 1958 were in complete or near-complete operation in the latter year (N. R. Chen, Table 3.42 and *Major Aspects of the Chinese Economy Through 1956*, p. 152). A substantial proportion of the 60 million ton increase in output during the First FYP came from new mines.

The claimed *increment* in output for the year 1958 was greater than the entire output in 1957, which is hardly credible. Of this, however, 44 million tons were stated to have come from primitive mines using little or no

TABLE III–12
OUTPUT OF MINERAL PRODUCTS IN CHINA,
1949 TO 1959
(thousands of tons)

Year	Coal	Crude petroleum	Iron ore
1949	32,430	121	589
1950	42,920	200	2,350
1951	53,090	305	2,703
1952	66,490	436	4,287
1953	69,680	622	5,821
1954	83,660	789	7,229
1955	98,300	966	9,597
1956	110,360	1,163	15,484
1957	130,000	1,458	n.a.
1958	270,000	2,264	n.a.
1959	347,800	3,700	n.a.

Sources:
N. R. Chen, 1967, Table 4.6.
Wu, 1963, p. 181.

machinery, and much of it had no commercial value. The actual level of production achieved in 1959, before the industrial collapse, still remains a matter of conjecture. It has been estimated, on the basis of the output of the major coal consuming industries, that the gross value of output of the modern mines increased by about 14 per cent in 1958 and by an additional 26 per cent the following year (Liu and Yeh, 1965, pp. 573 and 686). Translating the value data directly into tonnage would result in an estimate of 185 to 190 million tons in 1959, compared with Chinese official claims of 282 million tons from modern mines for the same year.

Although coal deposits have been found throughout China, production was concentrated in the industrial Northeast, about 50 per cent having originated in this area in 1952. Some 70 per cent of the major mine construction projects undertaken during the First FYP were located in the interior of the country, nearer the new industrial bases. Among them are the mining centers of Huainan, northeast of Wuhan, and Tatung, east of Pao-t'ou, for each of which output in excess of ten million tons was claimed in 1959 (Wang, 1960, p. 907). However, the major coal reserves are still in Manchuria, and it will not be easy for the Communists to alter the geographic distribution of production.

A rough estimate of Chinese coal production since 1959 was given in Table III–7. Most of the native mines started during the Great Leap Forward have been shut down (*Far Eastern Economic Review*, Nov. 14, 1963, p. 354). What has happened to the modern mines is more a function of demand than of supply. When operations in steelmaking and other heavy industries were curtailed after 1960, the demand for coal declined, for some 30 per cent of the coal mined in 1958 was consumed by this portion of industry (Wang, 1960,

p. 907). Household demand, which accounted for almost half the total output, would hardly have declined, however. An estimated production of more than 200 million tons in 1965 would still have left China far below the 550 million tons claimed by the Soviet Union, or the 400 million ton-plus level at which the United States has been operating, but it would make China the third largest coal producer in the world. In any event, the coal industry is not likely to constitute a bottleneck to future expansion of the Chinese economy.[12]

The situation with respect to iron ore is more ambiguous. There had been a consensus in the past that China lacked sufficient resources of high grade iron ore to support a large steel industry. Manchuria had about three-quarters of China's proven reserves, but their quality was low. The Communists claim, however, to have discovered large deposits in Northwest and Central China, with a workable reserve of five billion tons, enough to sustain a larger steel industry than is presently on the horizon. Output of iron ore rose from 4.3 million tons in 1952 to 15.5 million tons in 1956, but, despite this increase, shortages of ore have apparently led to occasional shutdowns of furnaces (*Major Aspects of the Chinese Economy Through 1956*, p. 47). Because of the low level of past industrialization, little scrap iron is available, so that the blast furnaces are dependent upon ore.

Securing an adequate supply of petroleum products has been a major problem for the Chinese, and they have been obliged to import substantial amounts from the Soviet Union. In 1959, reported domestic production was 3.7 million tons, while imports were on the order of 3 million tons, almost half in the form of gasoline (Eckstein, 1966, p. 110). A number of new refineries were built as part of the Soviet aid program during the First FYP.

Some extravagant claims with respect to oil reserves have been made by the Chinese government. A figure of six billion tons has been mentioned in recent years. If true, this would make China one of the principal oil-possessing nations in the world. A non-Communist estimate of 1.2 billion tons is more in line with what has been known of the industry (Chang, 1963, p. 7).

There is a good deal of difference between estimated reserves, proven reserves, and economically exploitable deposits, and this may account for some of the variation in the reserve figures. The two major fields actually in production, so far as is known, are those at Yumen in Kansu, which produced 44 per cent of the national output in 1958, and the recently discovered Dzungarian basin in Sinkiang.[13] Karamai, the central producing area of the latter field, grew from an uninhabited wilderness to a city of 30,000 during the decade following the discovery of oil in 1956. Both of these fields are located in the Far Western areas of the country, creating a need for the construction of pipelines and other transportation facilities.

12. This may not be true in the short run, however. As a result of the Cultural Revolution, the supply of coal in 1968 has been estimated at 20 to 25 per cent below the normal level. *New York Times,* July 14, 1968, p. 6.

13. A major strike was reported at Ta-ching in Northern China in 1964, alleged to have greater reserves than any field in China. The exact location of this field has never been disclosed. *Far Eastern Economic Review,* September 23, 1965, p. 565.

The Chinese claim to be no longer dependent upon foreign sources of petroleum, and the foreign trade data for recent years bear out this contention. This would mean that domestic crude production has caught up with refining capacity. The largest refinery in the country, that at Lanchow, with an annual capacity of a million tons, was built from 1955 to 1959 to utilize the output of the Yumen fields. The two other large refineries, at Shanghai and Dairen, were operating partly on the basis of imported crude and partly with Yumen crude shipped by rail and sea. Where these two refineries are now securing their crude oil is not known (Chang, 1963, pp. 18–20).

It was generally believed until recently that lack of domestic oil would be a major bottleneck to Chinese development. This view is no longer held as firmly, though Chinese production and consumption are still very small by Western standards.[14] Proven reserves now appear to be adequate to sustain a much larger industry. The Chinese problem seems to be that of investing in additional transportation facilities and refining capacity in order to exploit the new fields that have been brought in. A number of signs point to Chinese concern with this problem, among them the inclusion of petroleum in the list of priority industries following the Great Leap Forward, and negotiation with Italy for the purchase of a petroleum refinery in 1964 (Lewin, 1964, p. 57).

The Electrical Power Industry

The Chinese electric power industry at the inception of the First FYP was backward even in comparison with underdeveloped countries. In 1952, installed capacity of generating plants was 1.96 million kilowatts, compared with 2.3 million for India at the same time (N. R. Chen, 1967, Table 4.66 and Government of India, 1953, p. 78). On a per capita basis, the comparison was even less favorable for China.

The Chinese adopted Lenin's definition of Communism: "Soviet power plus electrification of the entire country." The First FYP included an ambitious program of capital construction which raised capacity 136 per cent and output 166 per cent in the course of five years. Over one hundred large units were begun, 24 of them part of the Soviet aid program. A doubling of capacity was claimed for the period 1957 to 1959, but it is subject to the same reservations as other Chinese statements about the Great Leap period (N. R. Chen, Tables 4.8, 4.65, and 4.66).

During the First FYP period, thermal plants received preference to hydro-electric projects because of the shorter construction time and the less complicated construction technology required for the former. This is not to say, however, that hydro-electricity was neglected. The large Fengman station in Manchuria, which had been badly damaged during the war, was rebuilt with

---

14. Annual per capita consumption was four to five gallons in 1963, compared with 130 gallons for Japan, 200 gallons for the USSR, and 900 gallons for the United States. *Far Eastern Economic Review,* October 7, 1965, p. 93.

Soviet aid and its capacity raised to almost 600,000 kilowatts, about one-third of the Grand Coulee project in the United States. A new station was begun at the Sanmen Gorge on the Yangtze River, with eight turbines of 137,000 kilowatts each, and was slated for completion in 1960; but, after the delivery of one generator, the Russians stopped further deliveries and it is not known whether the Chinese were able to build the equipment themselves (C. Y. Cheng, 1963, p. 47). From 1952 to 1957, the share of hydro-electric power capacity in total capacity more than doubled, but this was not reflected in the output figures, presumably because of the below-capacity operation of the new hydro facilities (N. R. Chen, Tables 4.66 and 4.8).

A rough estimate puts total generating capacity in 1964 at 14.16 million kilowatts; if true, it represents a substantial increase over the 9.4 million kilowatts claimed for 1959. Although hydro-electric capacity rose relative to the pre-Leap period, it still constituted only one-third of total capacity (Chao, 1967, p. 10).

More than any other industry, the production of electric power was concentrated in coastal areas, and particularly in the Northeast. Manchuria had 36 per cent of installed capacity in 1952, and because of the emphasis on speedy reconstruction of damaged facilities, its share rose to 41 per cent in 1957. (*Major Aspects of the Chinese Economy Through 1956*, p. 87). The large expansion planned for the second and third FYP's was calculated to alter the regional distribution radically, particularly in favor of the new Northwest industrial bases.

Among the deficiencies of the electric power industry, as catalogued by the Chinese themselves, were the following:

1. The largest thermal unit in China in the nineteen-fifties had a capacity of 50,000 kilowatts, one-fifth to one-tenth the capacity of generator units being installed in developed nations. The Chinese equipment, moreover, was of backward design with respect to temperature and pressure characteristics.

2. China has considerable water power potential in its narrow river gorges, but only a small fraction was developed.

3. The rate of coal use in terms of output of electricity was quite high (*Ibid.*, pp. 128–129).

Electric power, like petroleum, was accorded priority in the new order of things after the failure of the Great Leap Forward. Whether the technical deficiencies described in 1958 have been overcome remains an open question.[15]

Transportation

The Chinese Communists took over a country woefully deficient in modern transport. There were only 13,200 miles of railway track and 48,400 miles of surfaced roads in 1949. What this means can be judged by the fact that the continental United States, with a land area one-third smaller than China's,

---

15. Net production of electric power was estimated to be 37 billion kilowatt hours in 1965, compared with the previous peak of 44 billion in 1960. Ashton, 1967, p. 307.

had 397,000 miles of track and 1,865,000 miles of surfaced roads in the same year. This deficiency imposed an enormous initial handicap upon a government ambitious to create a modern industry.

Years of war and neglect had left even the existing facilities in very bad shape. Railroad beds and tracks were in a state of disrepair, bridges were down, while rolling stock, ships, and the few trucks available were in uniformly dilapidated condition. Every journey was an adventure, with arrival at one's destination by no means certain.

In the initial reconstruction stage, first priority was accorded the railroad network, three quarters of which was concentrated in Manchuria and the Northern provinces (Chang, 1961, p. 534). The length of trackage in operation rose from 22,000 kilometres in 1950 to 24,500 in 1952, of which about half represented new lines (N. R. Chen, 1967, Tables 3.44 and 6.1). Most of the investment went into restoration of existing facilities.

Some 8.5 per cent of total investment undertaken during the First FYP was allocated to railroad construction. There was an increase of 5,300 kilometres of operating lines between 1952 and 1959; all but 500 kilometres consisting of new lines were located mainly in the Northwest and Southwest, areas which had been virtually devoid of railway facilities (*Ibid.*) Among the most significant of the lines started during the period was extension of the line from Sian, in central China, through Mongolia and into Sinkiang, which will permit full utilization of the oil resources there, and will connect with the Soviet Turkestan railroad. Another major new line links Chungking with Chengtu in the province of Szechwan, and thence north to connect with the Sinkiang line, a loop which will help open up the agricultural hinterland of Szechwan.

The volume of traffic rose sharply during the First FYP. Ton-miles of freight carried increased by 123 per cent, and passenger-miles by 80 per cent (N. R. Chen, 1967, Table 6.4). This was achieved more through expansion of the volume of rolling stock than by increased operating efficiency. Such indicators as daily runs per locomotive and freight car, turn-around time of freight cars, stopping time per run, and train speed, showed little improvement. However, there was an increase of 20 per cent in average load carried, suggesting the availability of larger freight cars (*Ibid.*, Table 6.14).

Railroad statistics took a rapid upward turn in 1958 and 1959. It was reported, for example, that the volume of goods carried doubled during the two years, about the same increase as in the five preceding years. How much of this can be credited is another story.

The total length of railway lines in operation at the end of 1958 was given at 31,193 kilometres. By 1965, the total had grown only to 36,000 kilometres, and a large portion of the increment had probably been built in 1959 and 1960 (Chao, 1968b, p. 65). The main undertaking since 1958 has been completion of the line to Urumchi, the capital of Sinkiang.

The building of railroads would appear to be ideally suited to a labor-rich country with a large land area, and particularly to one which has shown no hesitation in mobilizing manpower for essential tasks in remote areas. With its poor road network and infant motor industry, China will be reliant for many years to come on railroads to move people and goods. Why the authorities failed to take advantage of the lull in industrial investment after 1959 to accelerate development of one of the most important elements in the nation's infrastructure, when so much could have been achieved with labor-intensive techniques, is not at all clear.

Highways have been distinctly secondary to railroads in freight transportation. Starting with 80.8 thousand kilometres of improved roads in 1949, the Chinese claim to have completed a total of 167 thousand kilometres by 1955. There are some discrepancies in the data which render this claim dubious. From 1950 to 1955, construction of highways was put at 71 thousand kilometres, of which 13.8 thousand kilometres were new. The remainder represented reconstruction and repair work (N. R. Chen, 1967, Tables 3.47 and 6.1). Even if the total mileage constructed is added to the initial mileage, on the ground that the repaired roads were impassable and therefore not included in the 1949 total, there are still 15 thousand kilometres not accounted for. (The balancing figure, and some of the reconstruction work as well, may represent upgrading of secondary roads to improved highway status).

Data for years after 1955 are not comparable with earlier figures because of inclusion of low grade as well as improved roads in the total. By a rough estimate, 215 thousand kilometres of improved highway were in operation by the end of 1958, which indicates extensive road building during the years 1955 to 1958 (Chao, 1968b, Table A-4).

Tonnage carried, rather than mileage, provides the best index of the economic importance of roads to China. In 1949, freight volume carried by motor vehicles amounted to 250 million ton-kilometres, compared with 18.4 billion for the railroads. By 1957, the highway tonnage had risen to 3.9 billion, with 135 billion for the railroads in the same year. Thus, while the rate of increase of highway tonnage surpassed that of the railroads, in the aggregate the railways remained far more important as freight carriers. Domestic production of trucks did not begin until 1956, and for the four years 1956–1959, total output was only about 40,000 units. China was estimated to have, in all, 250,000 to 300,000 trucks in 1965, plus 10,000 buses and 40,000 to 50,000 passenger cars (*Far Eastern Economic Review*, Feb. 17, 1966, p. 275).

Water transport, which has traditionally been important in China, was of considerably greater significance than highway transport during the 1950's. Ships and barges carried 34 billion ton-kilometres in 1957, about nine times the truck tonnage and a quarter of the railway tonnage. Dredging and canal building is indicated by the information that the length of inland waterways navigable by steamships rose from 24 thousand kilometres in 1949 to 40 thousand in 1958 (N. R. Chen, 1967, Table 6.8). The Chinese merchant fleet of 150 to 200 ships was approaching one million tons in 1965, plus 400 additional ships of 1.4 million tons on charter (*Far Eastern Economic Review*, March 3, 1966, p. 399).

Chinese civil aviation is still in its infancy. In 1958, the entire system produced only 109 million passenger-kilometres (N. R. Chen, 1967, Table 6.10), the equivalent of carrying 22,000 passengers across the United States, which is a small fraction of the monthly capacity of a single major United States carrier. The Chinese have built small twin-engine planes which can take eight to ten passengers, but the system still seems to be operating with an assortment of 40 to 50 Russian and British aircraft, none of very recent vintage.

Many underdeveloped countries use the backs of their people to carry a substantial proportion of goods transported. It has been estimated for 1933 China that traditional transport (all means not involving mechanical power) contributed a greater amount in value added than did modern transport, but that by 1957, the relationship was reversed (Liu and Yeh, 1965, p. 161). Full-time employment in the traditional transport sector fell from 3.5 million in 1952 to 2.5 million in 1957 (Emerson, 1965, p. 128), but recent travelers' reports emphasize the continued presence of human-powered vehicles in the cities, and it is hardly likely that rural transport has undergone any extensive mechanization.

Investment in transportation during the First FYP was low, relative both to China's needs and to what other underdeveloped countries have been putting into their transportation systems. It is not generally appreciated how large a contribution the provision of transportation makes to a modern economy. "One out of every five units of energy consumed in the United States goes toward moving people or materials. One-fifth of the final value of all the end products and services in the U.S. economy is accounted for by transportation expenditure" (Landsberg, 1964, p. 51). The modern transportation system of China contributed 4 per cent of net domestic product in 1957, while the addition of traditional transport raises the total to only 6.5 per cent (Liu and Yeh, 1965, p. 61. Transport equipment manufacturing would add very little to this total). Transportation will require very heavy outlays in future years if China is to attain the degree of mobility for people and goods essential to the efficient functioning of an industrial economy.

The Construction Industry[16]

Construction activity is basic to economic development. The growth of industry, the provision of homes, schools, hospitals, office buildings and shops, of roads and railroads, and not least, of agricultural infrastructure, depend upon this industry. A great deal of construction was carried on in China in conjunction with farming, but the non-rural sector of the industry expanded greatly after 1949.

Two estimates of the growth of construction output from 1950 to 1958

---

16. This section is based largely on Chao, 1968b.

are shown in Table III-13. Despite the many conjectural elements in the estimates, the discrepancies are not great. The estimates show, roughly, a tripling of construction output from 1952 to 1956. In this case, the subsequent 1958 spurt may have real significance, since the Great Leap Forward was accompanied by strenuous construction activity.

TABLE III-13
INDEXES OF CONSTRUCTION OUTPUT IN CHINA, 1950 TO 1958
(1952 = 100)

Year	Based on output components	Based on input components
1950	41.0	34.7
1951	72.9	61.1
1952	100.0	100.0
1953	142.6	154.2
1954	188.4	196.3
1955	185.2	210.1
1956	278.8	326.2
1958	457.9	485.1

Source:
Chao, 1968b, p. 57.

By the same token, the volume of construction work must have fallen greatly in 1960. The government announced a cutback in new basic construction in 1961 and 1962, and directed that emphasis be placed instead on repair work and completion of partly finished projects. After 1963, however, renewed investment activity in chemicals, tractors, synthetic fibers, and defense material must have increased the volume of new building, though, as already noted, this did not apply to railway construction, which remained in the doldrums.

Because of the labor-intensive character of construction work, employment provides another index of output. Total employment in the construction industry proper rose from 400,000 in 1950 to 2,950,000 in 1958.[17] During the 1950's, well over half the construction workers were kept on a permanent year-round status, but after 1960, considerable demobilization took place. Prior to 1960 there appears to have been a shortage of skilled labor, but the subsequent downturn in activity created a corresponding surplus of these workers.

Huge numbers of farmers were recruited for special projects in water conservation and irrigation during the agricultural off season. Some 150 million people were said to have been engaged in this type of work in 1958,

17. Productivity implications cannot be read out of a comparison of the employment figures with the output data in Table III-13, since the employment figures do not include farmers engaged in part-time construction work.

the peak year. Military conscripts were often employed in road building, particularly in the more remote areas of the country.

In organizing the construction industry, the Chinese set out to emulate the Russian example in mechanization. A good deal was invested in heavy cranes, excavating machines, concrete mixers, and the like. It might be thought that, in view of the relative amounts of labor and capital at their disposal, the Chinese would have been better advised to utilize more labor-intensive processes. A large aggregate labor reserve, however, is not inconsistent with difficulties in securing specific types of labor at the time and place needed. Where time is of the essence, mechanized construction may prove to be cheaper than manual labor. Advocacy of labor-intensive construction techniques also assumes that unemployed labor can be mobilized for work at subsistence wages. The Chinese have in fact differentiated in their food rationing in favor of heavy work, and it is at least questionable whether they could employ, side by side, premium-wage skilled workers and subsistence-minimum helpers.

These considerations apart, the mechanization of construction appears to have been poorly planned and executed. Machines were idled for long periods by inefficient project planning and shortages of skilled operators. Maintenance and repair facilities were inadequate. Perhaps in reaction to overly ambitious mechanization schemes during the early years of the First FYP, after 1956 mechanization was limited to large, complicated projects and those in sparsely populated areas.

The Chinese also began initially to emulate the Soviet Union in standardization and prefabrication of buildings. By 1960, prefabrication of concrete structural parts may have accounted for as much as 40 per cent of the total volume of reinforced concrete, indicating considerable progress along this line. This type of construction organization, which gives little scope to consumer choice and results in a highly uniform finished product, was adopted by the Russians to economize on skilled labor at the construction site, save on materials, and facilitate outdoor work in the winter months. Some of the same considerations account for its appeal to the Chinese.

Construction was the most rapidly growing sector in China during the 1950's. A large if not efficient organization was built rapidly from scratch, and some of its achievements were impressive. The shakedown after 1960, and the reduced volume of work, may have improved its performance. The stepup in training of architects and civil engineers must also be paying off, making available to the Chinese construction industry the type of skill for which it was dependent upon the Soviet Union prior to the break between the two countries.

Manchuria

Although a detailed analysis of the regional pattern of Chinese industrial development is beyond the scope of this volume, it is imperative that at least brief reference be made to Manchuria, the pre-Communist base of Chinese industry, and still the most developed part of the country. Manchuria owes its early development to a combination of good natural resources and easy access to the world outside China. Extensive mining of coal started at the beginning of the present century. The availability of iron ore facilitated the development of an iron and steel industry, mainly under Japanese control. In 1945, the large complex at Anshan had a capacity of 1.96 million tons of pig iron and 1.3 million tons of steel, representing respectively 68 and 65 per cent of the entire national capacity (C. S. Chen, 1964). The 1960 target for Anshan was 2.5 million tons of pig iron and 3.2 million tons of steel (Hsia, 1964, p. 130). Anshan had two small rolling mills in 1945, but they were partly dismantled by the Russians. After 1952, plants for the production of rails, girders, plates, seamless tube, and cold rolled plate were built there, but their capacity is not known.

A diversified machinery industry has been created around the iron and steel base. The largest automobile plant in the country, at Changchun, began producing trucks in 1956 and a Chinese-designed automobile in 1959. The Dairen Shipbuilding Works turned out ships as large as 10,000 tons in 1958, compared with a 6,000 ton maximum in pre-Communist China. The Harbin Steam Turbine works claimed to be producing thermal generators of 25,000 kilowatts and hydro-electric generators of 72,500 kilowatts by 1963. Harbin and Shenyang are the centers of the nation's machine tool industry. The construction of several large chemical plants at Kirin made this area the most important producer of chemicals in China.

In terms of its role in Chinese industry, Manchuria had become, by 1960 the counterpart of Pittsburgh, Detroit, and the industrial Midwest of the United States put together. The Liaoning Province (which includes Anshan, Shenyang, and Dairen) contributed 27 per cent of China's total machinery output in 1955, and 22 per cent of all metal products (N. R. Chen, 1967, Table 4.52). The establishment of new heavy industrial bases at Wuhan and Pao-t'ou were designed to reduce this industrial dominance, but the collapse of 1960 probably left the position of Manchuria unimpaired.

Conclusions

The workmanlike manner in which the Communists restored the battered economy of China in the first years of their rule, and the giant steps toward industrialization taken during the First FYP, convinced many contemporary observers that China would outpace Soviet Russia in a forced march from economic backwardness into modernity. A Western evaluation published in 1959 stated:

It is probable that the Chinese economy will continue to develop at an impressive, speed. The foundation of heavy industry has already been laid and is now very rapidly expanding. Because of the extremely backward state of the country and the low average standard of living, very large increases in industrial production and substantial improvements in the people's standard of life can be brought about by means of relatively small injections of capital and the application of modern industrial techniques (Hughes and Luard, 1959, p. 204).

There is no gainsaying the fact that the Chinese industrial achievement, up to the time of the ill-fated Great Leap Forward, was impressive. In a few short years, the sinews of modern industry had been laid, and China was on her way to becoming a great industrial power. This progress was achieved, moreover, by the efforts of the Chinese people themselves. That the Soviet Union sold industrial equipment and technical assistance to China should not obscure the fact that China paid for both goods and advice out of its own production, derived mainly from agriculture.

Whether the economic collapse of 1960 was an inevitable concomitant of the pace of industrialization from 1952 to 1959, or whether it could have been avoided if the government had maintained good relations with the Soviet Union and refrained from the hyper-activity of the Great Leap Forward, is a question that may be debated by historians for years to come. Failure to take into sufficient account the critical importance of the agricultural sector to the industrialization effort would have led to a downturn in any event. On the other hand, the Great Leap itself probably accentuated the agricultural difficulties, and it is at least possible that the Soviet Union would have seen China through the worst crop years if the Chinese had been willing to make the appropriate ideological obeisance.

Although the facts of Chinese economic development during the 1960's are known as yet only in outline, it is clear that the aftermath of the Great Leap Forward was a Great Depression comparable to that which had enveloped the Western World three decades earlier. Industrial investment declined to a low level; many factories closed down and dismissed their workers, and most others operated on a reduced schedule. The full magnitude of the downturn may never be known, for industrial administration was disorganized. It was three or four years before an upward industrial momentum could again be achieved, and this time the rate of growth appeared to be much less spectacular.

It can hardly be doubted that in the absence of involvement in a destructive war China will become a major industrial nation. But the much-publicized Great Leap goal of surpassing the industrial output of Great Britain by 1972, when it was expected that Chinese steel output would reach a level of 40 million tons annually, has been forgotten. The Chinese government has not abandoned industrialization as an eventual aim, but the slogan of "politics takes command" continues to frustrate efforts to achieve a normal path of development.

CHAPTER FOUR

Agriculture

The economic goal of the Chinese leadership is to transform an agricultural country into an industrial power. Attainment of this goal is critically dependent upon the ability of the peasants to produce a surplus of food and fibers over and above their own consumption to sustain labor engaged in industrial activities and provide raw materials for industrial production and exports. The Chinese government attempted to extract the surplus from the countryside through agricultural taxation and compulsory delivery of farm produce at nominal prices. But the low labor productivity in agriculture caused by inadequacy of capital and abundance of labor kept the agricultural surplus at a low level. In slack years an inadequate supply of agricultural produce frequently became a major bottleneck in the development of industry. The basic problem faced by the Chinese policy makers, thus, is not merely to convert an agricultural surplus into industrial capital but, more fundamentally, to create that surplus through increased agricultural output.

Although agriculture was not accorded priority in the scheme of development until three successive years (1959–61) of food crisis, the central role of agriculture in economic development has long been recognized by the Communist leaders. In his writings, Mao Tse-tung has repeatedly pointed out the importance of agricultural expansion in the process of industrialization. In his view, agricultural development entails two types of reform, institutional and technical. Under the conditions prevailing in China, he argued that institutional reform should be given higher priority and should precede technical reform. In a speech delivered in July, 1955, Mao predicted that institutional reform would be accomplished in ten years, and technical reform in 20 to 25 years (*Communist China, 1955–59*, 1962, p. 104). He held that "only when socialist transformation of the social-economic system is complete and when, in the technical field, all branches of production and places wherein work can be done by machinery are using it, will the social and economic appearance of China be radically changed" (*Ibid.*).

Mao's ideas about the role of agricultural reform in Chinese industrialization have become the Communist Party's guiding principle in formulating policy. Although later events proved that agricultural organization could undergo much swifter and more drastic changes than Mao foresaw in 1955, the policy of giving priority to institutional over technical reform has remained unchanged.

Organizational changes in Chinese agriculture will be discussed below. In this chapter we examine the measures taken by the Government to implement technical reform in agriculture, and their effects on agricultural production. The following topics will be discussed: (1) the agricultural contribution to economic growth; (2) development measures; and (3) current agricultural policy.

The Agricultural Contribution to Economic Growth

Agriculture contributes to economic development in many ways.[1] It provides food for the urban population, exports to exchange for capital goods, labor and capital for manufacturing, raw materials for industry, and markets for industrial products. In Kuznets' classification, these contributions may be grouped into the product type, the market type, and the factor type.[2] Since available data do not permit a discussion of all these types of contribution in the case of Chinese agriculture, only the product contribution will be examined. However, because the expansion of farm output has been the most important goal of agricultural policy, an analysis of the product contribution to a great extent reflects the performance of agriculture as viewed by the planners.

Trends in the gross output of agriculture are shown in Table IV-1. The large differences between the two estimates shown in the table are due primarily to two factors. First, certain biases in the official data were adjusted for by Liu and Yeh. Second, the coverage of the two series is not the same. For example, the value of farm subsidiary production, which is included in the official index, is omitted from the Liu-Yeh estimate. That there are no significant discrepancies between the gross and the net production indices in the Liu-Yeh series reflects the assumption of a constant ratio of cost to gross

---

1. For a general discussion of agriculture's contribution, see Johnston and Mellor, 1961, pp. 571–81.
2. Kuznets, 1961. According to Kuznets, "if agriculture itself grows, it makes a product contribution; if it trades with others, it renders a market contribution; if it transfers resources to other sectors, these resources being productive factors, it makes a factor contribution." Kuznets' article, together with comments by Ohkawa and others and Kuznets' reply, also appears in *Proceedings of the International Conference of Agriculture Economics*, London, 1963, pp. 39–82.

TABLE IV-1
AGRICULTURAL-PRODUCTION INDEXES FOR CHINA, 1949–57
(1952 = 100)

Year	Official index of gross value of agricultural production[a]	Liu-Yeh index of agricultural production[b]	
		Gross value	Net value
1949	67.3		
1950	79.7		
1951	86.7		
1952	100.0	100.0	100.0
1953	103.1	101.9	101.8
1954	106.6	104.1	103.8
1955	114.8	105.2	104.8
1956	120.4	108.2	108.1
1957	124.7	108.7	108.7

Sources:
[a] State Statistical Bureau, 1960, p. 118.
[b] Liu and Yeh, 1965, p. 140.

value of agricultural output. However, Communist data show that the proportion of cost to gross value increased slightly over the period, indicating a small increase in the "marketization" of the production process in agriculture.

The official index of gross agricultural production indicates that agricultural output rose very rapidly in the early phase of the Communist regime—48.5 per cent from 1949 to 1952.[3] The plausibility of the official claim cannot be properly evaluated, due to the lack of independent estimates for these years. Land reform was carried out rapidly and its effects on production are difficult to determine. The reform may have had favorable effects on production and productivity by increasing cultivated acreage and peasant incentive. Some previously uncultivated arable land, such as graveyards, was brought under plow.[4] Also as owners of the land, peasants found their income more directly related to their productive efforts. But the smaller size of farm units, the absence of government services to new owners to replace services rendered by former owners, and the countryside disorder which accompanied land-reform all tended to discourage production.

The net effect of land reform on agricultural production was probably unfavorable. On the other hand, after nearly 13 years of war and civil war, agricultural production in 1949-50 had dropped to perhaps two-thirds of the pre-war level. The restoration of political order, the maintenance of price stability, the reopening of urban-rural trade channels, the repair of

3. The gross value of agricultural production at 1952 prices was officially put at 32,590 million *yuan* in 1949 and 48,390 million *yuan* in 1952. State Statistical Bureau, 1960, p. 18.
4. The cultivated acreage increased from 97,881,000 hectares in 1949 to 107,919,000 hectares in 1952. N. R. Chen, 1967, Table 5.1. It is not clear, however, to what extent the increase in cultivated acreage can be attributed to land reform.

damaged irrigation facilities, and other efforts made by the Communists during 1949–52, undoubtedly contributed significantly to agricultural recovery. Their consequences may have far outweighed the unfavorable effects of land reform. If Communist statistics are to be credited, by 1952 the output of foodgrains had climbed 9.3 per cent above its pre-war peak level (Yang, 1956, p. 3). Compared with the 1949 levels, the total output and unit-area yields of foodgrains in 1952 were claimed to have increased by 44.8 per cent and 28.4 per cent respectively. Improvements were also recorded for nearly all other crops (*Ibid.*, p. 29). The probable accuracy of the Communist claims, in light of other, conflicting estimates, will be discussed below.

The claimed tempo of agricultural growth declined during the First FYP period (1952–57). The official index indicates that the gross value of agricultural production grew by less than 25 per cent during the period compared with an increase of 48.5 per cent during the three years from 1949 to 1952.

There is a consensus among Western experts that the Communist regime, in its early years, underestimated output in agriculture due to underreporting of production and landholding by peasants, as well as the lack of a well-established system of agricultural statistics. Underreporting gradually diminished in the process of agricultural socialization and became minimal by the end of 1956, when collectivization was nearly completed and statistical services were extended to rural areas. Liu and Yeh, in constructing their national income estimates for China, replaced official data on food crops for 1952–56 with their own estimate. When the Communist data were adjusted for conceptual differences, the Liu-Yeh estimate of agricultural production was greater than the Communist estimate for 1952 through 1956, but smaller for 1957 (Liu and Yeh, 1965, Table 68).

A more meaningful analysis of the contribution of agriculture to economic growth may be made in terms of the contribution to the growth of *total* product and of product *per worker*. Tables IV–2 and IV–3 present the results of applying Kuznets' formulae for computing the product contribution of agriculture to the Liu-Yeh data.

In Table IV–2 the rates of growth of total, agricultural, and non-agricultural production and the percentage share of the growth of agricultural product in the growth of total product are computed in both gross and net terms for the years 1953 through 1957, as well as for the entire period 1952–57. The agricultural contribution to the growth of net domestic product appears to have been greater than its contribution to gross domestic product, reflecting the smaller capital consumption allowances in agriculture than in the non-agricultural sectors.

For the First FYP period, the growth of agriculture accounted for only about 12 per cent of the growth of the total economy. This was largely the result of the policy of concentrating investment in industry. By the end of the First FYP, about 94 per cent of the growth of net domestic product and 95 per cent of the growth of gross domestic product were realized in non-agricultural sectors.

This disproportionate development is particularly evident when the relative contributions of the agricultural and non-agricultural sectors to the growth of national product per worker are compared. As may be seen from Table IV–3, for the first three years of the First FYP, agriculture actually made negative or zero contributions to the growth of domestic product per worker. Again this was partly the result of the neglect of agriculture in the distribution of state investment. In 1952, the agricultural product, which was smaller than the non-agricultural product, was produced by a labor force nearly three and a half times as large as the non-agricultural labor force. During 1952–57, the number of non-agricultural workers increased by less than five million persons, in contrast to an increase of about 16 million workers in agriculture.[5] In consequence, out of an increase of 65.2 *yuan* of national product per worker during 1952–57, only 1.1 *yuan*, or 1.7 per cent of the total, was accounted for by agriculture.

The failure of Chinese agricultural productivity to increase during the First FYP period parallels the Soviet experience. From 1928 to 1932, Soviet farm productivity may actually have fallen because of the destructive collectivization program undertaken under Stalin (Kershaw, 1953, p. 301). The Indian experience appears to have been similar to that of the Soviet Union: whereas food grain output rose from 50.8 to 82 million tons from 1950–51 to 1960–61, the portion of the economically active population engaged in agriculture almost doubled, from 71.8 million in 1951 to 137.5 million in 1961 (*Statistical Abstract of the Indian Union*, 1966, and ILO, 1957 and 1967). However, not too much weight should be placed on any of these productivity figures, since estimation of real agricultural labor inputs for such nations as China and India, as well as for Russia in the 1930's, is an exceedingly difficult proposition.

The contribution of agriculture to economic development can also be examined in terms of food supply. This analysis is particularly meaningful in view of the Chinese goals of autarky and self-sufficiency in grain production and because of the growing demand for grains caused by the industrialization program during the First FYP.

As noted above, Western experts believe that official data tended to understate actual grain production during the Communist regime. John Lossing Buck reached the following conclusion:

---

5. The Liu-Yeh data on output and employment in agricultural and non-agricultural sectors, 1952 and 1957 (in billions of 1952 *yuan* and millions of workers), are as follows:

	1952	1957
Agricultural		
Product	34.19	37.16
Employment	199.89	215.76
Non-agricultural		
Product	37.22	58.18
Employment	59.39	64.19

Source:
Liu and Yeh, 1965, pp. 66 and 69.

The 1949 amount was apparently based on the National Agricultural Research Bureau's estimates of production which were too low because the official land statistics of cultivated land recorded only three-fourths of all cultivated land. This omission of unregistered land was discovered in the study, *Land Utilization in China*. The Communists used a corrected figure for cultivated land for 1949, but somehow failed to make a corresponding correction in the amount of production. Hence, their series of production data record increases that are primarily statistical (Buck et al., 1966, p. 48).

Buck supports this conclusion with his own personal observation from many years of experience in China "that farmers continue to produce in spite of political and military upheavals, and that the areas seriously affected are small in relation to the total although the implementation of the violent land redistribution program during 1950–52 may have been more disruptive" (*Ibid.*, p. 28).

TABLE IV–2

THE RATIO OF THE GROWTH OF AGRICULTURAL PRODUCT TO THE GROWTH OF TOTAL PRODUCT IN CHINA, 1952–1957
(in per cent)

	1953	1954	1955	1956	1957	1952–57
I. Based on gross value product data						
A. Annual rate of growth						
1. Total	5.8	5.5	3.9	12.4	3.6	35.0
2. Agricultural	1.8	1.9	1.0	3.1	0.5	8.7
3. Non-agricultural	9.3	8.4	6.2	19.2	5.6	58.3
B. Contribution of agricultural growth to total growth	14.6	15.7	11.1	10.6	5.4	11.6
II. Based on net value product data						
A. Annual rate of growth						
1. Total	5.5	5.2	3.8	11.9	3.5	33.5
2. Agricultural	1.8	2.0	1.0	3.2	0.5	8.7
3. Non-agricultural	8.8	8.1	6.1	18.6	5.6	56.3
B. Contribution of agricultural growth to total growth	15.8	17.5	11.7	16.3	6.3	12.4

Note:

Figures in the table were obtained by applying Kuznets' formula for computing the product contribution of agriculture to the growth of national product to the Liu-Yeh findings on gross and net domestic product by industrial origin. (Liu and Yeh, 1965, Table 8).

The Kuznets formula is:

$$\frac{Pa \cdot Ra}{\triangle P} = \frac{1}{1 + \left(\frac{Pb}{Pa} \times \frac{Rb}{Ra}\right)}$$

where Pa = product of agriculture
Pb = product of all other sectors
$\triangle$P = increment in total product
Ra = rate of growth of Pa
Rb = rate of growth of Pb

Buck's opinion is consistent with the observation made by an Indian delegation sent to China in 1956 to study agricultural planning and techniques. According to the delegation report, "Chinese data after 1952 are not strictly comparable with earlier data. As such, a part of the improvement that is revealed by figures of area and yield of agricultural crops in China after 1952 over those of earlier years may be considered to be statistical" (Government of India, 1956, p. 86). The extent to which official data on grain production were understated for the early 1950's may be seen by comparing these data with the Dawson and Liu-Yeh estimates shown in Table IV–4. The

TABLE IV-3
THE RATIO OF THE GROWTH OF AGRICULTURAL NET PRODUCT PER WORKER TO THE GROWTH OF TOTAL NET PRODUCT PER WORKER IN CHINA, 1952–1957
(in *yuan* and per cent)

	1953	1954	1955	1956	1957	1952-57
I. Increase in national product per worker (1952 *yuan*)	10.9	−9.9	−7.3	29.3	7.8	65.2
1. Agricultural contribution	0	−0.4	−0.3	1.5	0.2	0.9
2. Non-agricultural contribution	12.1	12.8	7.0	27.3	3.4	64.0
3. Joint contribution	−1.2	−2.4	0.6	0.5	2.4	0.3
II. Total agricultural contribution to increase in product per worker (1952 *yuan*)	−0.6	−1.6	0	1.7	1.4	1.1
III. Relative contribution of agriculture to increase in product per worker (per cent)	−5.2	−13.9	0	5.8	17.9	1.7

Note:
The figures in this table are obtained by applying Kuznets' formula for computing the product contribution of agriculture to the growth of total product per worker to the Liu-Yeh data on employment and net domestic product, by industrial origin. (Liu and Yeh, 1965, Tables 8 and 11.)

The growth of national product per worker is the sum of (1) the increment in product per worker in agriculture, weighted by the share of agriculture in labor force at the end of the period; (2) the increment in product per worker in the non-agricultural sector, weighted by the share of non-agricultural sector in labor force at the end of the period; and (3) the change in the share of the non-agricultural sector in the labor force during the period, weighted by the difference between product per worker in the non-agricultural and agricultural sectors at the beginning of the period. Algebraically, the change in total product per worker is as follows:

$$\frac{P^1}{L^1} - \frac{P^o}{L_o} = \left(\frac{P_a^1}{L_a^1} - \frac{P_a^o}{L_a^o}\right)\frac{L_a^1}{L^1} + \left(\frac{P_b^1}{L_b^1} - \frac{P_b^o}{L_b^o}\right)\frac{L_b^1}{L^1} + \left(\frac{P_b^o}{L_b^o} - \frac{P_a^o}{L_a^o}\right)\left(\frac{L_b^1}{L^1} - \frac{L_b^o}{L^o}\right)$$

where
L_a = workers in the agricultural sector
L_b = workers in all other sectors
L = all workers = $L_a + L_b$
P_a = product of agriculture
P_b = product of all other sectors
P = total product = $P_a + P_b$
Superscripts o and 1 refer to time periods

TABLE IV-4
ESTIMATES OF GRAIN PRODUCTION IN CHINA, 1949–1957

	1949	1950	1951	1952	1953	1954	1955	1956	1957
1. Total output (millions of tons)									
a. Official	108.1	124.7	135.1	154.4	156.9	160.5	174.8	182.5	185.0
b. O. L. Dawson	150.0			170.0	166.0	170.0	185.0	175–180	185.0
c. Liu-Yeh				176.7	180.4	184.5	186.4	182.3	185.0
2. Output per capita (kilograms)									
a. Official	202	228	242	272	270	262	288	294	290
b. O. L. Dawson	275			296	284	282	301	278–286	287
c. Liu-Yeh				306	311	310	307	297	290

Sources:
(a) N. R. Chen, 1967, pp. 127 and 338.
(b) E. F. Jones, 1967b, p. 93.
(c) Liu and Yeh, 1965, pp. 54, 102, and 132.

difference between these two independent estimates and the official estimate diminishes over time and disappears by 1957.

If Dawson's estimate is accepted, half the claimed increase in grain output during 1952–57 was statistical. If the Liu-Yeh estimate is accepted, more than two-thirds of the reported increase for 1952–57 represented an exaggeration. In per capita terms, grain production declined during 1952–57 according to the independent estimates, while official data show a slight improvement. In any case, all estimates, both official and independent, indicate that per capita output of grain in 1957, the last year of the First FYP, was no greater than 290 kilograms. Not all of this was available for direct human consumption. During 1952–57, from 16 to 22 per cent of grain output in China was diverted into other uses, such as feeding animals, seeding, industrial processing, and exports (Liu and Yeh, 1965, p. 132). The amount available for per capita consumption was thus in the neighborhood of 250 kilograms, the traditional Chinese subsistence standard (Fei, 1945, p. 158).

Toward the end of the First FYP, therefore, the Chinese planners faced serious difficulties which were not encountered in the Soviet Union. In 1932, the last year of the first Soviet FYP, grain output per capita was 458 kilograms despite the agricultural disruption accompanying forced collectivization (Yeh, 1967, p. 343). At the end of their respective first planning periods, the Chinese people were close to the margin of subsistence, while there was still considerable room for raising further the rate of saving from Soviet agriculture.

During the first Chinese FYP, agricultural raw materials constituted over 50 per cent of the gross value of industrial output, and about 85 per cent of the raw materials used in consumer goods industries came from agriculture (Chi, 1957, pp. 4–8). Poor harvests had adverse effects upon industrial production in general and on light industry in particular. This may be seen from the data on harvest conditions and annual rates of industrial growth during 1952–57, shown in Table IV-5. Years of good harvests were followed

TABLE IV-5

HARVEST CONDITIONS AND ANNUAL RATES OF INDUSTRIAL GROWTH IN CHINA, 1952–1957

	1952	1953	1954	1955	1956	1957
Harvest conditions	Good	Average	Poor	Good	Poor	Average
Annual rate of growth (per cent)						
1. Gross value of total factory output		33.0	13.8	6.7	39.7	8.7
2. Gross value of factory output of consumer goods		28.8	14.9	−2.9	31.5	−1.3
3. Gross value of handicraft output		8.9	10.4	−3.8	20.0	−3.3

Source:
The annual rates of growth are computed from the Liu-Yeh estimate of the gross value of factory and handicraft output at 1952 prices. Liu and Yeh, 1965, Tables 42 and 43.

by high rates of industrial growth, years of poor harvests by low rates. These relationships were particularly evident for factory output of consumer goods and for handicraft output. Thus, good harvest conditions in 1952 and 1955 were largely responsible for the relatively high rates of growth of factory output of consumer goods and of handicraft output in 1953 and 1956, and bad weather in 1954 and 1956 caused a decline in those outputs in the following years, 1955 and 1957.

The dependence of industry upon agriculture for the supply of raw materials may be judged further in terms of the production of industrial crops, of which cotton is undoubtedly the most important. Table IV–6 presents data on the

TABLE IV-6
THE SUPPLY OF COTTON AND THE PRODUCTION AND SALES OF COTTON TEXTILE PRODUCTS, 1952–1957

	1952	1953	1954	1955	1956	1957
1. Supply of cotton (thousand tons)						
a. Production	1,304	1,175	1,065	1,518	1,445	1,640
b. Import	77	11	42	86	44	35
c. Total supply	1,381	1,186	1,107	1,604	1,489	1,675
2. Production of cotton textile products						
a. Cotton yarn (thousand bales)	3,618	4,104	4,598	3,968	5,246	
b. Cotton cloth (million metres)	3,829	4,685	5,230	4,361	5,803	
3. Rate of capacity utilization (per cent)						
a. Cotton yarn		82.2	87.9	90.3	76.6	94.0
b. Cotton cloth		86.7	91.8	89.8	71.7	89.4
4. Retail sales						
a. Cotton cloth (million metres)	2,881	3,999	3,888	3,936	5,281	

Sources:
Department of Industrial Statistics, 1958, pp. 169 and 182.
Ma, 1958, p. 32.
State Statistical Bureau, 1960, p. 119.

production and import of cotton, the rate of utilization of spinning and weaving equipment, and the production and retail sales of cotton textile products during 1952–57.

Cotton production fluctuated during 1952–57 not only because of weather conditions but due to government policies as well. Output of cotton declined in both 1953 and 1954, while cotton imports in these two years also fell. The adverse effects of the decline in cotton output upon the cotton textile industry were strongly felt in 1955, when the production of both cotton yarn and cloth fell to more than 15 per cent below the previous year's level. The rates of capacity utilization for both yarn and cloth also declined. But a bumper harvest and increased imports raised the level of cotton supply in

1955, resulting in an increase in the output of cotton yarn and cotton cloth in 1956. In the same year, the rate of capacity utilization rose to 94 per cent for cotton yarn and 89.4 per cent for cotton cloth. Again, poor harvests and reduced imports of cotton in 1956 caused the output of cotton textile products and the rate of capacity utilization to decline in 1957.

A discussion of the contribution of Chinese agriculture to economic development would not be complete without a few words about farm subsidiary activities, which include thousands of occupations ranging from hog and poultry raising and vegetable growing on small private plots to household processing of agricultural products and the manufacture of small farm implements.

Farm subsidiary occupations in China, as in many other underdeveloped agricultural countries, traditionally provided idle time employment opportunities for farmers and for members of their families who were otherwise unproductive. During the First FYP, farm subsidiary production accounted for as much as 30 per cent of the gross value of agricultural output (Perkins, 1966, p. 72). Official data indicate rural processing activities alone accounted for about 20 per cent of the gross value of agricultural output (N. R. Chen, 1967, Table 5.93). A significant portion of the products of farm subsidiary occupations went for urban consumption and for export. In addition, even at the height of agricultural collectivization in 1956 and 1657, "the private plot might be very important in raising the calorie supply well above bare subsistence,"[6] and "supplied most of the fertilizers used by Chinese agriculture, in the form of pig manure."[7]

Agricultural cooperation caused the output of subsidiary products to decline, particularly in 1956, when the cooperative cadres encroached heavily on individual independence in production. For example, in various areas of Szechwan Province the output of subsidiary products fell from 5 to 15 per cent in 1956 (Perkins, 1966, p. 73). The loss of 17 million pigs between July, 1954, and July ,1956, resulted not only in serious shortages of pork and pork products but also in a loss of 510,000 tons of ammonium sulphate, or 4.6 kilograms per hectare of arable land (Walker, 1965, p. 56).

Agricultural Production During and After the Great Leap Forward

The Communist policy of concentrating investment in industry to the neglect of agriculture, combined with the fact that the population was growing more rapidly than the agricultural output available for food consumption, led to

6. Walker, 1965, p. 32. Walker estimated that in 1956, 14 per cent of the total calorie supply of Chinese peasants came from the private plot. More important was the fact that the private plot "provided the kind of foods needed for a balanced diet" and that "in many cases it raised the standard of living from the margin of subsistence to tolerable levels." *Ibid.,* p. 41.

7. *Ibid.,* p. 43. On July 1, 1956, approximately 70 million out of a total of 84.4 million pigs in China were privately owned. These 70 million pigs provided 2,064 million tons of ammonium sulphate equivalent per year, or 18.7 kilograms per arable hectare.

the inability of agriculture to support industrial development. Toward the end of the First FYP, agricultural stagnation became a critical bottleneck to further growth of industry. It was against this background that the Great Leap strategy was adopted in December, 1957.

Since the start of the Great Leap Forward, official grain statistics have been published only for 1958 and 1959. But Communist officials have discussed their grain estimates with foreign visitors. In consequence, an official estimate can be constructed for the years beginning in 1960. However, Western experts differ about the validity of the official statements, and a number of independent estimates have been made.

The official and independent estimates are presented in Table IV-7. The

TABLE IV-7

ESTIMATES OF GRAIN OUTPUT IN CHINA, 1957–1967
(in million tons)

Year	Official-Constructed[a]	O. L. Dawson[b]	U.S. Agricultural Officer in Hong Kong[c]	E. F. Jones[d]	W. Klatt
1957	185	185	185.0	185	
1958	250	204	193.5		
1959	270	170	167.7		
1960	150	160	159.0		150[e]
1961	162	170	166.5	162	
1962	174	180	178.3		
1963	183	185	179.1		
1964	200	195	182.7		
1965	200	193–200	179.9	200	
1966	200[f]	193.5[g]	175.0[h]		180[i]
1967			190.0[j]		

Sources:
[a] These data are constructed by E. F. Jones, 1967b, p. 93.
[b] These estimates are cited by E. F. Jones, *ibid.*
[c] Reports of the U.S. Agricultural Officer, American Consulate General's Office, Hong Kong, cited in Emery, 1966.
[d] E. F. Jones, *ibid.*
[e] W. Klatt, 1961, pp. 64–75.
[f] Constructed by Dwight H. Perkins, 1967, p. 36.
[g] E. F. Jones, 1967a, p. 236.
[h] Editor, *Current Scene,* Nov. 10, 1966, p. 12.
[i] W. Klatt, 1967, p. 154.
[j] *New York Times,* July 14, 1968, p. 6.

official estimates were constructed by Jones for 1958–65, and by Perkins for 1966. The estimates of the U.S. Agricultural Officer in Hong Kong are prepared "with the use of the files which have recorded the soils and cropping practices in the various farm districts of China" (E. F. Jones, 1967a, p. 236). Dawson's estimates are based on the calculation of the optimum possible effect of new inputs of irrigation and fertilizer on the 1957 production base, which is then adjusted for "inefficient application of the new inputs and for offsetting impairments to farm resources" (*Ibid.*) Jones' estimates are derived from "a model incorporating bits of official information" (1967b, p. 94). Klatt employs the grain balance technique to obtain his figures (1961, pp. 64–75, and 1967, pp. 151–155).

There is a consensus among Western experts that the official claim of 250 million tons of grain output in 1958 was grossly exaggerated, although opinions differ with respect to the extent to which the data should be discounted. There is also agreement that grain output in China actually declined in 1959 despite official claim to the contrary. All estimates, including the official one, point to a sharp decline in 1960 grain production, and a gradual rise thereafter until 1964, a year of bumper harvest. But agreement ends there.

The debate largely centers on the plausibility of the estimates prepared by the U.S. Agricultural Officer in Hong Kong. The Hong Kong estimates are consistently lower than both the official and other independent estimates for the years since 1961. In 1965, for example, grain output was 200 million tons on the basis of both the official and the Jones estimates, and in a range of 193–200 million tons according to Dawson, while the Hong Kong estimate places it at 180 million tons, five million tons below the 1957 level. Grain output in 1966 was estimated by the U.S. Agricultural Officer at 175 million tons in contrast to an estimate of 200 million tons by Perkins, 193.5 million tons by Dawson, and 180 million tons by Klatt.

A proponent of the Hong Kong estimate argues that the unfavorable factors in agricultural production during the Great Leap, such as deterioration in the quality of land and the loss of animals, which resulted in a loss of at least 10 per cent (18.5 million tons) of the 1957 output, had not disappeared by 1965; and that the increased use of chemical fertilizers at best may have compensated for these ill effects by 1964–65, so that the level of output was restored approximately to the 1957 level. In this view, the assumption that grain output in 1965 was 15 million tons larger than in 1957 (as implied in both the official and the Jones' estimates) is untenable (Liu, 1967, p. 40).

On the other hand, several scholars are of the opinion that the Hong Kong estimate is implausible. Eckstein says, "if food crop production in 1965–1966 was only 175 and 180 million tons as compared to 185 million in 1957, one would expect a significant decline in per capita consumption of foodstuffs, given the continued growth in population between 1957 and 1966" ("A Letter from Professor Alexander Eckstein . . . ," 1967, p. 239). This does not appear to be consistent with travelers' and refugee reports which "project an image of a reasonably comfortable food situation with no indications of food stringency" (*Ibid.*).

Jones is suspicious of the Hong Kong estimate because it is based on a formula using sparse current local information, which "may be primarily recording past performance and current crop weather and be missing new secular trends in farm output resulting from changes in irrigation and fertilizer supply" (Jones, 1967a, p. 236). His choice of a higher estimate "rests on the recognition of a strong tendency towards a stable relationship between a population and its food supply" (*Ibid.*). In his view, such a relationship has been maintained in China since 1964.

Perkins argues in similar fashion that "the Hong Kong data... indicate China's development of chemical fertilizers and other modern farm inputs has had virtually no effect on grain output" (1967, p. 37). "The implications of the Hong Kong data," he continues, "point up the essential implausibility of the estimates. A variety of materials all suggest that per capita rations have risen substantially since 1961"(*Ibid.*, p. 38). He concludes that "the grain estimates reconstructed from statements by Chinese officials plausibly reflect what we know of the performance of agriculture in China since the disasters of 1959–61" (*Ibid.*, p. 39).

A critical element in the evaluation of grain output is population. Most Western scholars accept the 1953 census as a lower limit and assume that the Chinese population grew at an annual rate of slightly more than 2 per cent in the 1950's, and from 1.5 to 2 per cent in the 1960's. This would yield population totals between 735 and 755 million for 1966. (See Chapter V for further discussion of population totals). If these figures are plausible, it is difficult to accept a grain estimate for 1965–66 as low as 175–180 million tons, for it would mean that from 1957 to 1965–66, the population would have increased by 90–110 millions, while at the same time output fell by 5–10 million tons. Per capita supply of grain would have declined from 286 kilograms in 1957 to 238–249 kilograms in 1966 (Perkins, 1967, p. 37).

Moreover, not all grain produced is available for direct human consumption. As noted above, from 16 to 22 per cent of the grain output during the First FYP was diverted to other uses. Even if we assume a lower percentage of grain output for purposes other than direct human consumption, the per capita consumption of grain would still be much lower than 250 kilograms, the traditional Chinese subsistence standard. The caloric intake implied in the amount available for per capita consumption would suggest a diet below the minimum subsistence level. (This point is considered further in reference to living standards in Chapter VII). However, if one accepts a lower population estimate,[8] and assumes that Chinese population in 1966 was no higher than 700 million, grain output of 175–180 tons would not have been out of the question.

We cannot leave the subject of grain production without discussing imports. Beginning in 1961, China imported annually between four and six million tons of wheat and barley from Canada, Australia, and Argentina, costing approximately $300–500 million per year (MacDougall, Jan. 27, 1966, pp. 121–125, and Wilson, May 20, 1965, pp. 352–54). Agricultural recovery since 1962 did not change the policy of importing wheat from the West; net

---

8. Contrary to the opinion of most Western demographers, Werner Klatt believes that the population count of 1953 was overstated. Moreover, he uses a very low rate of population growth between 1960–61 and 1966–67. According to his estimates, Chinese population rose from 650 million in 1960–61 to 700 million in 1966–67, or at less than 1.3 per cent per year.

grain imports still amounted to 5.5 million tons in 1965 and five million tons in 1966. (Price, 1967, p. 601, and Editor, *Current Scene*, Nov. 10, 1966, p. 12). Continued importation of wheat in recent years has led some to doubt that food output has regained the 1957 level.

There are indications, however, that China is importing wheat for valid economic reasons, for rice is being exported, as it was even during the crisis years of 1959–61.[9] This policy of substituting imported wheat for Chinese rice has probably yielded a comparative advantage to China. Rice is a labor-intensive crop and does not easily lend itself to mechanization, while wheat is ideally suited to large-scale mechanized farming. In China, moreover, rice is generally grown on the fertile land of the South and Southwestern provinces while wheat is found in the less productive regions of the North and Northeast. As a result, the unit yield of rice in China is generally higher than that of wheat.

China probably cannot compete with the West in wheat production at current market prices; ton for ton, rice should bring a much higher financial yield for China than wheat (*China Report*, Dec. 1965—Jan. 1966, pp. 4–5). Moreover, it is probably cheaper to feed China's urban population on the East coast with imported wheat from foreign countries than to produce wheat in the North and Northeast and ship it by an inadequate transport system. Some of the land freed from wheat production can be used for the production of rice. There was a Chinese report in 1963 that rice growing had been extended into the wheat region of North China.[10]

While we know little about Chinese grain output in recent years, our knowledge of the production of industrial crops is even poorer. Like food crops, the output of economic crops declined sharply during the crisis years 1959–61. The available evidence seems to indicate that the recovery of economic crops did not occur until 1963, whereas grain production took an upturn two years earlier (Emery, 1966). Before 1963 the Chinese were preoccupied with the problem of avoiding starvation and concentrated their efforts on expanding grain production. Economic crops did not receive much attention until the outlook for grain production became better in 1962.

The most important economic crop in China is cotton. Table IV–8 presents the official claims for cotton output for 1957–66, as reconstructed by Edwin F. Jones, together with Jones' own estimate and the estimate of the U.S. Agricultural Officer in Hong Kong. The official figures for 1958–59 are considered by Western experts to be greatly exaggerated, although the extent of the inflation in cotton output data may have not been as great as for grain.

Cotton production probably began to drop in 1959, although official statistics register an increase for that year. The decline continued until 1962,

9. In 1965, for example, China exported 810,000 tons of rice to Albania, 300,000 tons to Japan, 180,000 tons to Cuba, Malaysia, Ceylon, Hong Kong and other countries. *Far East Economic Review*, Sept. 30, 1965; and *Studies in the Problems of Mainland China (Ta-lu wen-ti chuan-ti yen-chu)*, Jan 31, 1966.

10. *Far Eastern Economic Review, 1964 Yearbook*, p. 136. For a discussion of the possibility, see Ishikawa, 1967, pp. 67 *ff.*).

TABLE IV-8
ESTIMATES OF COTTON PRODUCTION IN CHINA, 1957–1966
(in thousand tons)

Year	Official claims[a]	Hong Kong estimate[b]	Jones estimate[c]
1957	1,640	n.a.	1,640
1958	2,100	1,900	
1959	2,410	1,800	
1960	n.a.	1,400	
1961	n.a.	900	900
1962	850	900	
1963	1,020	1,000	
1964	1,410	1,200	
1965	1,650	1,300	1,400
1966	1,700		

Sources:
[a] N. R. Chen, 1967, pp. 338-9; E. F. Jones, 1967a, p. 238.
[b] Reports of U.S. Agricultural Officer, American Consulate General's Office, Hong Kong (cited in Emery, 1966).
[c] E. F. Jones, 1967b, p. 94.

when total production was only 55 per cent of the 1957 level. Information supplied by emigrants asserts cotton production in 1962 fell as low as 850,000 tons on almost 2.5 million hectares (Jones, 1967a, p. 238). Output began to recover in 1963, thanks to an increase in sown acreage (670,000 hectares) and to the use of more fertilizers and insecticide, better irrigation, and higher quality strains (*Far Eastern Economic Review, 1964 Yearbook*, p. 136). Peking claimed a 20 per cent increase in output and acreage in 1963.

Output was officially reported to have risen by 37 per cent in 1964, average yield by 23 per cent, and sown acreage by 11 per cent (Jones, *ibid.*). The increase in acreage was partly due to the extension of cotton growing areas southward to the Yangtse Valley, where the climate is said to be better suited to cotton.[11] The U.S. Agricultural Officer in Hong Kong doubted the official claim. The Hong Kong estimate places Chinese cotton output in 1964 at 1.2 million tons, an increase of 20 per cent over the 1963 level but 210,000 tons below the official estimate. On the other hand, Jones believes that "the crash expansion of the cotton textile industry in 1965 after the 1964 cotton harvest lends credence to the 1964 cotton production claim" (Jones, 1967a, p. 238).

According to official reports, cotton production continued to rise substantially in 1965 and slightly in 1966. Total output and unit yields in 1965 were at the highest level ever recorded in China, according to statements at the Fifth National Conference on Cotton Production, convened in Peking in February, 1966 (*Ta-Kung Pao*, Feb. 9, 1966, pp. 1–2). Western observers, however, were generally skeptical. Jones felt that "yield and output claims

---

11. A *People's Daily* article of March 4, 1965 stated that cotton yields in South China were almost double those in North China.

seem inconsistent with supply indications and continuing policies" (Jones, 1967b, p. 94). His estimate and the Hong Kong data put cotton production in 1965 at 1.4 and 1.3 million tons, respectively, in contrast to the official claim of 1.65 million tons. But there seems to be no question that 1965 was the third good year in succession for the cotton harvest, although the picture for 1966 was less clear.

The best performance of Chinese agriculture after the Great Leap Forward was probably in subsidiary production. Farm subsidiary products constituted a significant portion of the gross value of agricultural output during the First FYP. But with the advent of the commune system, private plots were confiscated and rural free markets abolished. The output of family sideline occupations dropped sharply, as did subsidiary products produced collectively in the communes. Although aggregate statistics are not available to substantiate this, an index may be found in the number of pigs. Since pigs provide manure, meat, bristles, hides, and hog casings, all of them exportable commodities, they were a principal item among subsidiaries. When communization of pigs was instituted in 1958 and 1959, poor management of the commune rearing units brought on disease and a decline in the number of pigs in many places in 1960. The death rate among young pigs was over 50 per cent (Walker, 1965, p. 84).

Agricultural failures forced the regime to reopen rural markets in a few areas in late 1959, and to return private plots to the peasants in the summer of 1960. Beginning in 1961, special efforts were made to activate rural trade fairs as a more positive incentive to subsidiary production. Various enterprises and institutions were encouraged to raise their own subsidiary foods. Complete self-sufficiency in vegetables was said to have been achieved by the Army, for example (Perkins, 1966, p. 92). After the spring of 1961, the policy for pig-breeding was "simultaneous development of private and collective, with emphasis on private." The policy was so effective that Vice Premier T'an Chen-lin claimed a record number of pigs reared in China at the end of 1964 (*China News Analysis*, Oct. 1, 1965, p. 3). But observers in Hong Kong believed that by the end of 1965, the pig population had recovered only to the level of 1957, when it was officially put at 145.9 million (P. H. M. Jones, Dec. 23, 1965, pp. 566–67).

Since 1964, the Chinese have stressed diversification of farm subsidiary activities. Agricultural diversification is not new to China. It was common in traditional China and received some attention by the Communists during the 1950's. Only since 1964, however, has an all-out effort been made by the regime to encourage the peasants to engage in a wide variety of farm activities. About 20 per cent of the commune labor force has been organized into teams devoted to subsidiary production. In areas where the potential for subsidiary activities is great, over half the peasants may be diverted into these activities.

The results have been quite impressive. An exhibition of agricultural subsidiaries held in Peking in February, 1966, showed that in 1964 and 1965, several thousand kinds of products were turned out in eleven main categories, including crop growing, animal raising, weaving and spinning, fishing, hunting, processing, and transporting (*People's Daily*, Feb. 16, 1966, p. 3). Subsidiary production has become a major source of cash income for the peasants, and in many communes contributes more than half of their gross income. For the country as a whole, subsidiaries are said to have constituted one-third of the gross value of agricultural and subsidiary products and 40 per cent of the state procurement of these products (*Ibid.*).

While subsidiary production provides a major source of the peasants' income and makes a worthwhile contribution to their diet, its most important contribution, from the government's viewpoint, is the enlargement of the export trade. There has been a substantial increase in the Chinese export of subsidiary products in recent years. Their percentage share in total exports to most of the countries available for comparison also increased.[12]

Two points should be noted with regard to subsidiary production in China. First, its scope has apparently undergone substantial changes in recent years. During the First FYP, the subsidiary work of farm households was limited to four types of activities: (1) hunting, fishing, and the collection of natural products, such as herbs, fuel, and minerals; (2) processing of own farm crops, milling rice, grinding flour, ginning cotton, and shelling ground nuts; (3) handicraft work for own use, such as tailoring and shoe-making; and (4) handicraft work done for other consumers who supply the raw materials (Li, 1958, p. 56). Because the definition of subsidiary production has now been expanded to include certain kinds of work not previously covered, such as crop growing, animal raising, and transportation, the distinction between subsidiary and principal occupations is not always clear. For example, cotton and oil seeds grown on small patches of land are sometimes counted as subsidiary crops (MacDougall, April 21, 1966, pp. 153–56, and *China News Analysis*, Oct. 1, 1965).

Secondly it is difficult to determine the extent to which an increase in subsidiary output may be attributed to family or to collective sideline occupations. No information has been made available on the proportion of family subsidiary output in total subsidiary output. One study indicates that 80 per cent of the pigs and 95 per cent of the poultry are raised on private plots, which represent perhaps 5 to 7 per cent of the total cultivated land (P. H. M. Jones, Dec. 23, 1965, pp. 566–67). On the other hand, collective sidelines have been reported quite frequently in the Chinese press, where they are called the only correct road to socialist production.

Agricultural Development Policy

The performance of Chinese agriculture since the Communist takeover may be summarized as follows: agricultural production had regained the pre-war

---

12. Hong Kong was an exception. But in aggregate terms, the import of subsidiaries from China to Hong Kong increased from H.K. $995 million in 1964 to H.K. $1,162 million in 1965. MacDougall, April 21, 1966, pp. 153–156.

Agriculture

level by 1952, and continued to improve slowly until the Great Leap Forward of 1958–59, which was followed by three consecutive years of crop failures. Recovery began for food grains in 1962, and for industrial crops in 1963. By 1967, agricultural production as a whole probably exceeded the 1957 level.

To what extent was production affected by organizational change? As one reviews the brief history of Chinese agriculture under Communism, one gains the impression that, while land reform and collectivization may have contributed to output during the years prior to 1957, the commune system aggravated, if it did not directly cause, the difficulties of 1959–61, and that the relaxation of the commune program made an important contribution to the agricultural recovery after 1962. In the long run, however, the level of agricultural production is determined by technique. In evaluating the past performance of Chinese agriculture and its future prospects, it is useful to examine the measures that have been adopted to expand and utilize efficiently the available resources.

Land is the most important factor in an overpopulated country like China. Production may be raised by increasing either the area of cultivated land or the level of land productivity. The total cultivated acreage of mainland China was put at 245 million acres in 1949; it reached a peak in 1956–57, when it was 276 million acres, and then fell by ten million acres in 1958 and 1959 (N. R. Chen, Table 5.1, and *China Pictorial*, Feb. 1961, pp. 2–3). The present level of cultivated acreage is probably no higher than the 1956-57 level.[13] Cultivated area per capita was 0.47 acres in 1952, declined to 0.45 acres in 1957, and has fallen further in recent years. At the present time, the per capita acreage is probably less than 0.4 acres[14] compared with 2.96 acres for the Soviet Union, 3.08 acres for the United States, and 8.4 acres for Canada (Hsaio Yu, Feb. 9, 1958, pp. 21–24).

The area of reclaimable wasteland in China is said to be about 250 million acres, nearly equal to the present cultivated area and largely scattered in Sinkiang, Heilungkiang, and the Southern provinces (Kwangtung, Kwangsi, Yunnan, and Fukien). A survey was made of 140 million acres of the wasteland, of which about 80 million acres were believed worth cultivating. During the First FYP, 12.8 million acres were reported reclaimed.[15] This figure, however, did not represent the net increase in cultivated acreage, for official statistics indicate that the cultivated acreage increased by only 9.7 million acres from 1952 to 1957. This suggests that, during the First FYP, about 3.1 million acres of cultivated land were allocated to such non-agricultural uses

13. In 1966, Peking claimed that the area of cultivated land was 1.6 billion mou, or 267 million acres, which is a reduction of 4.6 per cent from the peak of 276.3 million acres in 1957. See Larson, 1967, pp. 250–251.

14. Assuming that total cultivated acreage at present is the same as in 1957, i.e., 276.3 million acres, and total population is 750 million, the cultivated area per capita would be 0.37 acres.

15. Hsaio Yu, Feb. 9, 1958, pp. 21–24. According to Larson, the acreage reclaimed during 1953–57 was also variously reported as 13.1 million acres and 9.6 million acres. Larson, *ibid*. But, as indicated in the text, the latter figure seems to be the increase in cultivated acreage, which is not the same as the acreage of reclaimed land.

as the building of industrial plants, railways, highways, airfields, and large dams and reservoirs.[16] Moreover, the reportedly reclaimed area of 12.8 million acres included five million acres of landholdings which were concealed from the authorities in the early years of the regime and were uncovered during the agricultural collectivization that began in the autumn of 1955. (Hsaio Yu, *ibid.*). Consequently, the area actually reclaimed from 1952 to 1957 was 7.8 million acres, only 3 per cent of the total area which was believed arable but remained uncultivated.

Of this relatively small reclaimed area, approximately 3.3 million acres was the result of peasant effort to cultivate small patches of wasteland near village sites, including graveyards, roads, footpaths and boundaries. The remaining 4.5 million acres were reclaimed by state farms (3.4 million) and resettlements (1.1 million, Hsaio Yu, *ibid.*).

The small role of the state in land reclamation was due to the limited amount of state investment allocated to agriculture and to the high cost of reclaiming wasteland. It was estimated that reclamation of one acre of wasteland in China cost an average of at least 360 *yuan*. To bring under cultivation the 80 million acres of arable land which were considered worth reclaiming would have cost about 30 billion *yuan*, equivalent to nearly 70 per cent of state investment in agriculture during the entire period of the First FYP (Hsaio Yu, *ibid.*). Thus, as Chou En-lai told an Indian agricultural delegation in 1956, high costs limited the scope of land reclamation (Government of India, 1956, p. 171). Moreover, reclamation without careful planning might also hasten soil erosion.

Since 1957 the Chinese do not appear to have made any major efforts to expand their reclamation program.[17] They have depended primarily on intensive cultivation to raise land productivity. Multiple cropping, a traditional Chinese method for the intensive exploitation of land, has been given a high priority by the Communist authorities in their efforts to raise agricultural production. The multiple cropping index, i.e., the ratio of sown acreage to cultivated acreage, rose from 130.9 per cent in 1952 to 145 per cent in 1958 (State Statistical Bureau, 1960, p. 128). The index for China may be compared with 195 per cent for Taiwan, 138 per cent for Japan, and 103 per cent for India.[18] During the First FYP, 72 per cent of the increase in grain production was said to have been the result of multiple cropping (Hsaio Yu, Sept. 9, 1957, pp. 5–8).

During the agricultural crisis of 1959–61, natural calamities and mismanagement of the commune program caused both cultivated and sown

16. This was confirmed by Li Fu-ch'un, Chairman of the State Planning Commission, in November, 1957, p. 106.

17. Reclamation in 1959–61 was primarily an attempt to reclaim land which was deserted during the Great Leap. (See Larson, *ibid.*) However, a large area of wasteland was reportedly opened up by a number of mechanized state farms in Heilungkiang in 1963. P. H. M. Jones, Nov. 12, 1964, pp. 350–352.

18. The figures for Taiwan and Japan are for 1956. Dawson, in Buck et al., 1966, p. 134. The figure for India was for the mid-1950's; Government of India, 1956, p. 156.

acreage to decline, and probably resulted also in a decline in the multiple cropping index. Table IV-9 contains official and Western estimates of sown acreage for selected years from 1952 to 1965, from which it appears that sown acreage in 1965 had regained the 1957 level. The multiple cropping index, however, probably surpassed the 1957 index of 140.6 per cent. One estimate puts the index at 150 in 1966.[19]

TABLE IV-9
ESTIMATE OF SOWN ACREAGE IN CHINA, 1952–1965
(million hectares)

	1952 (official)	1957 (official)	1961	1963	1964	1965
1. Food grain crops						
a. FEER	112.3	120.9	—	118.7	120.0	120.5
b. Jones	112.3	120.9	120.0	—	—	125.0
2. Soybeans						
a. FEER	11.7	12.7	—	8.0	8.1	8.1
b. Jones	11.7	12.7	7.0	—	—	9.0
3. Cotton						
a. FEER	5.6	5.8	—	4.2	4.5	5.0
b. Jones	5.6	5.8	3.5	—	—	4.8
4. Oilseeds (peanuts, rape, sesame)						
a. FEER	4.7	5.8	—	3.6	4.1	—
b. Jones	4.7	5.8	3.5	—	—	4.5

Sources:
Official data for 1952 and 1957: N. R. Chen, 1967, pp. 286–87.
The *FEER* (*Far Eastern Economic Review*) estimate: Munthe-Kaas, Feb. 3, 1966, pp. 153–155; P. H. M. Jones, Sept. 30, 1965, pp. 613–615; and *Far Eastern Economic Review, 1965 Yearbook*, p. 130.
The Jones estimate: E. F. Jones, 1967, p. 94.

A second method of raising land productivity is to increase unit area yields. Table IV-10 compares Chinese yields per unit of sown acreage for major crops with those of other countries. Chinese yields were substantially lower than those in developed nations, but compared favorably with yields in the underdeveloped nations. When one looks at the data, the same familiar story emerges: progress from 1952 to 1957, decline in 1961, then a gradual improvement from 1962 to 1965, when yields appear to have surpassed the 1957 level.

The Chinese leaders were well aware that although agrarian reorganization may have been desirable for the creation of conditions conducive to agricultural development, in the final analysis increases in output would have to be attained primarily through technical reform. In the first decade of the

---
19. This figure can be derived from estimates of chemical fertilizers used per acre of cultivated area and of sown area. Larson, 1967, p. 246. According to E. F. Jones, the index stood at 143.1 in 1965. 1967b, p. 94.

TABLE IV-10

A COMPARISON OF AVERAGE CROP YIELDS IN CHINA WITH SELECTED COUNTRIES, 1952 TO 1957
(in kilograms per hectare and per cent)

Crop	Average yield in China (kilograms per hectare)	CHINESE AVERAGE YIELD AS PERCENTAGE OF THAT OF								
		Australia	Burma	Canada	Egypt	Phillipines	India	Japan	Pakistan	U.S.	U.S.S.R.
Foodgrains											
Rice	2,539	46.0	173.0	96.0	57.7	208.5	186.7	42.8	183.7	77.0	
Wheat	823	72.8		56.8	55.4		196.5	61.4	109.5	80.6	
Coarse grains	1,013			66.9	37.6		113.0	38.7		63.6	
Potatoes	1,903			55.2	46.5		212.0	50.2		49.8	
Soybeans	795			53.2	49.3		111.0	48.9		47.2	
	1,434							69.5		57.7	
Economic Crops											
Cotton	239		177.8		47.5		228.3		108.1	59.0	29.4 (1956)
Jute	2,085						177.4		120.0		
Tobacco	976	114.5	114.0	66.0		182.5	140.0	56.9	88.3	65.3	
Peanuts	1,233	120.0	214.2		67.6		170.3	88.5	115.2	118.9	
Rapeseeds	461			54.0			121.4	37.9			
Sesame	374		32.0				181.4		88.4		
Sugar cane	38,923	74.3			49.2	81.5	120.8	121.4	74.9		
Sugar beets	11,983			46.9				56.8		32.8	70.3 (1953-56)

Source:
Statistics for China computed from data given in Ministry of Agriculture, 1958.
Statistics for other countries were drawn from the yearbooks of the Food and Agricultural Organization of the United Nations.

Communist regime, efforts were made to implement a number of technical improvement measures. In addition to the extension of multiple cropping areas, these included water conservation, more intensive use of manures and fertilizers, soil conservation and improvement, use of better seed, more high-yielding crops, pest control, better farm implements, and the gradual introduction of mechanized farming (State Planning Commission, 1956, pp. 83–87). These reforms were popularized through emulative drives organized by party workers and through technique popularization stations. In the mid-1950's, there were over 10,000 such stations, one for every twenty villages, and the number was said to be growing rapidly (Government of India, 1956, pp. 141–46). Various technical measures essential to the improvement of yield were spelled out in detail. In 1958, they were summarized by Mao in a famous "eight-character charter,"[20] representing eight different methods of increasing agricultural production.

The crop failures of 1959–61 brought home the realization that the nation's further economic development depended on the ability to obtain steadily rising agricultural yields. The creation of stable, high-yielding farms throughout China has become the overriding goal of farm policy. While the eight-character charter is still frequently cited as a means of attaining this goal, special emphasis has been laid on water conservation, fertilization, and farm mechanization, in that order.

The frequent occurrence of floods and droughts is one of the tragic phenomena of Chinese history. In a period of 2,142 years, from 206 B.C. to 1936, 1,035 droughts and 1,037 floods were recorded (Teng, 1958, p. 48). In the absence of effective control measures, natural calamities resulted in great damage to crops. In 1928, for example, droughts in thirteen provinces caused agricultural output to decline to 20 per cent of the normal level (Wang, March 20, 1965, pp. 40–46). In the four years from 1953 to 1956, natural calamities affected more than 51 million hectares of cultivated land, resulting in a loss of about 37.5 million tons of foodgrains, an average of 9.4 million tons a year. During the same period, the loss of cotton amounted to 410,000 tons, averaging 125,000 tons a year, which would have been sufficient to produce cotton cloth for the needs of more than 70 million persons for a year (Hsiao Yu, Sept. 9, 1957, pp. 5–8).

Without effective control of water, both in preventing flooding and water logging and in storing against drought, the level of agricultural production in China will continue to depend on nature. Drought and flood apart, lack of irrigation facilities on a large portion of the cultivated land was also responsible for low yields. According to one estimate, if irrigation were available, the yield of foodgrains from current dry fields could be increased by at least 50 per cent in the North and 100 per cent in the South. Expansion of irrigated acreage in the first eight years of the Communist rule was said

20. The eight characters are: *shui* (water conservation); *fei* (fertilization); *t'u* (soil conservation and improvement); *chung* (seed selection); *mi* (close planting); *pao* (plant protection); *kung* (tool improvement and innovation); and *kuan* (farm management).

to have led to an increase of 12.5 million tons of foodgrains (Lo and Shangkuan, Oct. 9, 1957, pp. 15–17).

Water conservation is not new to China. For more than twenty centuries the peasants maintained soil fertility by irrigation from rivers and reservoirs. But the large scale on which the Communists mobilized the peasants to build water conservation projects was unprecedented not only in Chinese, but also in human, history. According to Communist statistics, irrigated area rose from 16 million hectares in 1949, to 21.3 million hectares in 1952, representing respectively, 16.3 per cent and 19.8 per cent of total cultivated area. There was a further expansion of 13.3 million hectares from 1952 to 1957, which would have put 31 per cent of the cultivated land under irrigation at the end of the First FYP. The irrigated area allegedly doubled in 1958, the year of the Great Leap, to 66.7 million hectares, or 62 per cent of total cultivated area. In contrast to the increase of 1958, no significant expansion was reported for 1959–60.[21]

Chinese irrigation statistics must be viewed with caution, since they do not necessarily indicate actual improvement. Much of the alleged increase during 1953–57 represents statistical improvement resulting from correction for understated amounts in earlier years (Dawson in Buck et al., 1966, pp. 149–167). This was particularly evident in 1956, when an expansion of irrigated acreage of 7.3 million hectares, an increase of nearly 30 per cent over 1955, was recorded (N. R. Chen, 1967, p. 289).

Moreover, a number of official statements testify to the fact that a large portion of the irrigated area was not able to withstand drought. In 1957, for example, the Vice Minister of Water Conservation reported that of the 34.7 hectares of land under irrigation in that year, only 10.1 million hectares were capable of resisting drought (Buck et al., 1966, pp. 149–167). With respect to the large number of water conservation projects built in 1958, a Russian expert noted that "because of improper management, many projects met with accidents, damaging the materials of the people and the state," (*Ibid.*). In fact, only 74 per cent of the area reportedly under irrigation as of April 30, 1958, was actually irrigated.[22]

According to Vice Premier T'an Chen-lin, out of a total of 67.1 million hectares of irrigated area in 1959, only 40.7 million hectares were capable of resisting a drought lasting from 30 to 70 days (1960). An editorial in *People's Daily* in June, 1959, complained that some reservoirs had no water, others contained water but no aqueducts, and that some of the water works could not be utilized. The situation was worsened by the breakdown of irrigation equipment in 1959–61. It was admitted in the Chinese press that in these years repair work could not keep up with deterioration and failure (Chi, 1965, pp. 37–54).

21. Claimed irrigated acreage was 67.1 million hectares in 1959. T'an, 1959, p. 2. It was 66.7 million hectares in 1960. Chi Ssu, 1960, pp. 2–4.
22. *People's Daily*, May 3, 1958. As of April 30, 1958, of the total irrigated area, 75 per cent in the North, 67 per cent in the Northeast, 72 per cent in the Northwest, 79 per cent in the East, 69 per cent in the Southwest, and 74 per cent in the Central South were effectively receiving sufficient water.

The effective use of irrigation projects remained an important problem when agriculture began to improve. Toward the end of 1963, only 60 per cent of the irrigation systems were functioning (*People's Daily*, Nov. 30, 1963, p. 1). In 1964, 30 to 40 per cent of the large and medium-sized irrigated projects were not effectively linked up (*People's Daily*, Feb. 4, 1965, p. 2).

Because of the ineffectiveness of the systems, irrigated acreage has been scaled down to a considerable degree. One source puts the total at about 33 million hectares at the end of 1964 (*Economic Bulletin*, Nov. 16, 1964). Another indicates that about 1.3 million hectares were added during 1965 (*People's Daily*, Sept. 30, 1965). These indicate a total of 34.3 million hectares of irrigated land at the end of 1965, or 31.6 per cent of the total cultivated area, about the same percentage as in 1957 (Lin Pin, 1966, pp. 1–6).

During the First FYP, only 2,550 million *yuan* were invested in water conservation, representing 4.6 per cent of total state investment. Moreover, much of the investment was spent on major river projects, leaving a far smaller amount available for farm irrigation. The state encouraged peasants to build small irrigation projects on their own. It was said that for large-scale projects, the irrigation of an additional *mou* would cost the state 32 *yuan*, and for medium-sized projects, 17 *yuan*. On the other hand, an additional *mou* of irrigated area obtained through small projects carried out by the peasants involved a subsidy of only 0.69 *yuan* plus a loan of 4 *yuan*. Furthermore, large-scale projects required a construction period of two to four years, and medium-sized projects, one year, while small projects could usually be completed in a single season.

Some 21 large-scale irrigation projects were initiated during the First FYP, but few of them had been completed by the end of the period. Of the expansion in irrigated acreage, 90.8 per cent came from the peasants' small projects (Lo and Shang-kuan, 1957, pp. 15–17). They involved the construction of canals, dams, ditches, storage pools and pumps, the digging of wells and ponds, and the harnessing of small rivers, carried out with almost bare hands and little technical guidance. Many of the small projects proved unable to resist drought.

Despite the poor performance of small irrigation projects, the peasants were mobilized to build them on a huge scale during the Great Leap. The construction of major reservoirs, which had siphoned off a large portion of state investment in water conservation during the First FYP, came to a halt. In August 1958, the Central Committee of the Communist Party issued a "Directive on Water Conservancy Work" which stressed "Three Primary Principles": (1) emphasizing the building of small projects; (2) concentrating on the storage rather than on the diversion of water; and (3) relying on the communes rather than on the state for construction (*People's Daily*, Sept. 11, 1958).

At the same time the major emphasis was shifted from anti-flood drainage operations to the use of water resources to combat drought. The storage of water is particularly important for certain regions in China, such as those north of the Tsin Divide, where the rainfall is generally limited to the summer season and where the flow of rivers is not adequate to the needs of irrigation on the adjacent farmland (Elliot, 1965, pp. 50–51).

The "Three Primary Principles" remain the basic official policy on water conservation. Since 1961, however, greater emphasis has been placed on improving and repairing the reservoirs already built, linking up existing irrigation systems, and strengthening the management of irrigation work (Liu Sheng, 1965, pp. 21–26). An editorial in the *People's Daily* stated that the main task of water conservation during the Third Five Year Plan (1966–70) "will not be the building of a large number of 'backbone' projects, but efforts will be concentrated on extending irrigation and drainage systems and bringing the potential of existing projects into full use" (Quoted in Munthe-Kaas, 1966, p. 153). This does not mean that the state will not invest in large projects. In fact, some large-scale projects are under construction, such as the Pi-Shih-Hang Canal in Anhwei, the new outlet to the sea from the Heilungkiang River, and the gigantic network of power transmission lines and electric pumping stations on the Yangtse and Pearl River Deltas and in the Tungting Lake area. But the main effort will apparently be concentrated on small local works because "such construction requires much less outlay in investment than the construction of large or medium-sized water engineering work, and shows quicker results." (*People's Daily*, Aug. 5, 1964, p. 1).

The current water conservation campaign is operating on three levels. First, the state is building a number of large projects. As of early 1965, there were more than 60 large-scale reservoirs built by the state, all of them having storage capacities of more than 100 million cubic meters (Elliot, 1965, pp. 50–51). A number of large hydroelectric power stations also are being built. At the next level, communes are being organized to combine their efforts with those of the state in the construction of irrigation systems, such as digging canals and building electric pumping stations in areas including the Pearl River delta, the T'ai-hu and Tungting Lake areas, the Western Sezchwan plain, and the Yangtse River Valley. Finally, and most important, countless small works are being built with local resources by the peasants themselves. They are building small reservoirs and digging branch canals and ditches to link local farms up with large reservoirs and the main canals. The reservoirs, with capacities ranging in capacity from 100,000 to one million cubic meters, are reported to run into the tens of thousands (*Ibid.*).

At all three levels, wherever possible, labor-intensive methods are employed. The basic policy is "to utilize as fully as possible the vast human resources in China and to substitute labor for capital through labor accumulation in order to make up insufficiency of capital funds." In accord with this policy not only small works but large projects as well are to be built "in the spirit of the mass line" (*People's Daily*, April 11, 1965, p. 2).

With a sufficient supply of water, the increased use of fertilizers, particularly chemical fertilizers, is the most effective way of raising unit yields.

It has been estimated that production in the post-war years could have been raised 25 per cent above the pre-war level with the use of 15 million tons of chemical fertilizers to supplement local manures (Shen, 1951). Experiments and soil analysis indicated that Chinese soils were more responsive to nitrogen than to phosphate, and more to phosphate than to potash.[23] On the average, the application of a unit of nitrogenous fertilizers would raise the output of foodgrains by three units or that of cottonseeds by one unit, while the use of a unit of phosphorus fertilizers would increase the output of foodgrains and cottonseeds by only 1.5 units and 0.5 units respectively.[24] To raise crop production, China will have to use a sufficient amount of nitrogenous fertilizers.

Since 1949, efforts have been made to utilize every possible source of manure and to introduce the use of chemical fertilizers, particularly nitrogenous fertilizers. For centuries the Chinese farmer has preserved soil fertility through the use of such organic manures as night soil, stable manure, compost, green manure, crop residues, mud matter, and oil cakes. In the early years of the Communist regime, due to the absence of a chemical fertilizer industry, government measures were aimed at fuller and more scientific use of the manure resources of the villages. In the early 1950's, it was estimated that about 85 per cent of the total cultivated area was manured through organic matters, of which 50 per cent was night soil and stable manure, 20 to 30 per cent compost, and 10 to 15 per cent green manure (Government of India, 1956, p. 146). These organic fertilizers were rich in nitrogen, phosphorus and potash. One estimate put the availability of plant nutrients from organic fertilizers in 1956 as follows: nitrogen, 3,809,000 metric tons; phosphate, 1,741,000 metric tons; and potassium, 3,235,000 metric tons. In terms of kilograms per hectare of cultivated land, the figures were 34 for nitrogen, 15.5 for phosphate, and 29.3 for potassium (Dawson in Buck, et al., 1966, pp. 118–19). Another estimate indicates that pig manure alone contributed 24 kilograms of ammonium sulphate to each hectare of arable land in the mid-1950's (Walker, 1965, p. 56).

Reliance on organic manures was not sufficient to raise unit yields from the level traditionally obtained to that required by the Communist development program. "Increase in organic fertilizer is to a large extent a function of the increase not only in human and animal population but crop residues as well.

---

23. Chang and Richardson in their field experiments for widely scattered localities in 14 Chinese provinces beginning in 1935 found that 74 per cent of the soil tested was deficient in nitrogen, 38 per cent lacked phosphate, and only 12 per cent needed potash. Other estimates indicated that nitrogen deficiency was found in 80 to 96 per cent of the total cultivated land, phosphate in 40 to 55 per cent, and potash in 15 to 24 per cent. See Lamer, 1957, pp. 426–427; and *China News Analysis,* February 15, 1963.

24. Wang and Han, 1957, pp. 11–15. According to this source, experiments made during 1935–42 for 152 localities in fourteen provinces including Shantung, Honan, Hupeh, Kiangsu, Yunnan, Kweichow, and Shensi suggested that on the average, the use of a unit of ammonium sulphate would increase the output of paddy rice by 3.6 units and that of cottonseeds by 0.7 units, and that a unit of super-phosphate applied to the soil would raise the output of both wheat and rapeseeds by 1.3 units.

TABLE IV-11
CONSUMPTION OF CHEMICAL FERTILIZERS IN CHINA, 1952–67

Year	Production (000 tons) (1)	Imports (000 tons) (2)	Consumption (000 tons) (3)	Cultivated area (million hectares) (4)	Sown area (million hectares) (5)	Consumption per hectare of cultivated area (kilograms) (6)	Consumption per hectare of sown area (kilograms) (7)
1952	194	239	433	108	141	4.01 (0.80)	3.00 (0.60)
1953	249	343	592	108	144	5.48 (1.10)	4.11 (0.82)
1954	321	579	900	109	148	8.26 (1.65)	6.07 (1.21)
1955	345	875	1,220	110	151	11.10 (2.22)	8.08 (1.62)
1956	663	837	1,500	112	159	13.39 (2.68)	9.24 (1.85)
1957	803	997	1,800	112	157	16.07 (3.21)	11.46 (2.25)
1958	1,244	1,456	2,700	108	156	25.00 (5.00)	17.34 (3.47)
1959	1,765	1,190	2,955	108	156	27.36 (5.47)	18.94 (3.79)
1960	2,460	860	3,320	107	142	30.74 (6.15)	23.37 (4.65)
1961	1,400	883	2,283	107	142	21.33 (4.26)	16.08 (3.22)
1962	2,100	1,000	3,100	107	156	28.97 (5.77)	19.87 (3.99)
1963	2,900	1,700	4,600	107	156	42.99 (8.60)	29.42 (5.88)
1964	3,500	1,030	4,530	109	156	41.56 (8.31)	29.03 (5.80)
1965	4,500	2,500	7,000	109	156	64.22 (12.84)	44.87 (8.97)
1966	5,000	3,500	8,500	101	151	84.15 (16.83)	56.28 (11.26)
1967	6,000	3,500	9,500	101	151	94.06 (18.81)	62.91 (12.58)

Note:

Figures in parentheses are measured in terms of plant nutrients. The gross consumption is converted into plant nutrients at a ratio of 5 to 1. See J. C. Liu, 1965a and b.

Sources:

(1) 1952–66: These estimates are based on sources in the U.S. Department of Agriculture, and are given in Larson, 1967, p. 246.
1967: E. F. Jones, 1967a, pp. 237–8.
   (2) 1952–66: Larson, *ibid.*
1967: The amount of fertilizer imports in 1967 is assumed to be the same as the 1966 level. Writing in late 1966, Larson noted that negotiations for imports in 1967 about equal to those in 1966 already had been completed. (*Ibid.*)
   (3) The sum of production and imports.
   (4) 1952–58: N. R. Chen, 1967, pp. 284–5.
1959–60: *China Pictorial*, February, 1961, pp. 2–3.
1961: E. F. Jones, 1967b, p. 94.
1962: Derived from the Larson estimates of total availability of chemical fertilizers and the amount per unit of cultivated area for 1962. Larson, *ibid*.
1963: Assumed to be the same as the 1962 level.
1964–65: The 1965 figure is estimated in E. F. Jones, 1967b. The 1964 figure is assumed to be the same as that of 1965.
1966: Derived from the Larson estimates of total availability of chemical fertilizers and the amount per unit of cultivated area for 1966. Larson, *ibid*.
1967: Assumed to be the same as the 1966 level.
   (5) 1952–58: N. R. Chen, *ibid.*
1959: The sown acreage is assumed to remain at the 1958 level.
1960–61: The 1961 figure is estimated. E. F. Jones, 1967b. The 1960 figure is assumed to be the same as that for 1961.
1962–65: The 1965 figure is Jones' estimate. The 1962 figure is Larson's estimate. The sown acreage in 1963 and 1964 is assumed to remain unchanged between 1962 and 1965.
1966: Derived from Larson's estimates of total availability of chemical fertilizers and the amount per unit of sown area for 1966.
1967: Assumed to be the same as the 1966 level.
   (6) Derived by dividing total consumption by total cultivated area.
   (7) Derived by dividing total consumption by total sown area.

But within such a close cycle it is not possible to meet the need for more crop nutrients without bringing in substantially more chemical fertilizer" (Dawson in Buck et al., 1966, p. 134).

Chemical fertilizers had never been used in China in any appreciable amount. There were only two chemical-fertilizer plants, the Dairen Plant in Manchuria and the Yung-li Plant in Nanking, with annual productive capacities of 230 and 35 thousand metric tons of ammonium sulphate respectively. Total output reached a peak of 277 thousand metric tons in 1941, but declined to 27,000 metric tons in 1949. (J. C. Liu, 1965, pp. 28–52). There were some imports, but the quantities were small.

The Communists have made strenuous efforts to increase the supply. Production and imports of chemical fertilizers, and their consumption per hectare of both cultivated and sown areas during 1952–67, are shown in Table IV-11. Total production was reported to have exceeded the pre-war peak level by 1953. During the First FYP, the productive capacities of the Dairen and Yung-li Plants were doubled, and eleven new plants were added to produce not just ammonium sulphate, but also ammonium nitrate and superphosphate (Yi Tseng, 1966, pp. 53–70). By the end of the plan, total output reached more than 800,000 metric tons. At the same time, imports increased steadily from 343,000 tons in 1953 to nearly one million tons in 1957. Except for 1961, the output of chemical fertilizers has shown a steady rise. Western experts estimate China's fertilizer output in terms of standard fertilizer at six million tons in 1967 (E. F. Jones, 1967a, pp. 237–8). Imports reached a peak of perhaps 3.5 million tons, making the total supply of chemical fertilizers in 1967 about 9.5 million tons.

In terms of plant nutrients as well, the consumption of chemical fertilizers per hectare of both cultivated and sown area has risen steadily, except for 1961. Nevertheless, Chinese consumption per hectare of cultivated area has not reached the world average. It is far below the 217.4 kilograms of plant nutrients for Taiwan, 304.5 kilograms for Japan, and 166.3 kilograms for Korea.[25] The use of organic manures in China provided 78.8 kilograms of plant nutrients per hectare of cultivated land in recent years (Dawson in Buck et al., 1966). The amount of plant nutrients per hectare of cultivated land derived from the use of both chemical and organic fertilizers is probably no more than 100 kilograms.

Thus, in spite of the rapid increase in the supply of chemical fertilizers in the past sixteen years, the current level of consumption relative to the size of cultivated acreage is far from sufficient. The crucial questions are, how much chemical fertilizer China requires, and what the prospects are for meeting these requirements.

Assuming that the Chinese population will continue to grow at a rate of 2 per cent per annum and that the sown acreage will remain at the level

---

25. The world average was 26.8 kilograms of plant nutrients in 1963–64. These figures refer to the consumption of chemical fertilizers during the 1963–64 fertilizer year. FAO, 1966a, p. 23.

reached in 1957, China will have to supply chemical fertilizers in the amount of 12 million metric tons to maintain per capita grain consumption in 1972 at the 1957 level of 260 kilograms (J. C. Liu, 1965, pp. 915–32). Another projection indicates that if per capita grain consumption is to attain 300 kilograms in 1972, given allowances for increased yields due to improved techniques and the use of organic fertilizers, a minimum of 15 million metric tons of chemical fertilizers will be required. This estimate is also based on the assumption that the population will continue to grow at an average rate of 2 per cent into the 1970's, and that food crop acreage can be increased by a net of ten million hectares from 1962 to 1972 (Dawson in Buck et al., 1966).

These conclusions are strikingly similar despite differing methodology. Both suggest that even allowing for improvements in agricultural technique and in the supply of other farm inputs, China will have almost to double the present chemical fertilizer output in order to meet grain requirements if the current rate of population growth continues into the 1970's.[26]

The experience of Japan and Taiwan indicates that to develop the chemical fertilizer industry and to distribute and utilize its products effectively will require a long period of time. In the 1890's, for example, the yield of paddy rice in Japan was about the same as the present yield in China, 2,500 kilograms per hectare. It took Japan more than half a century to develop a going program of fertilizer use. In 1962, seven million tons of chemical fertilizers were used in Japan on about eight million crop hectares, and the yield of paddy rice, 4,500 kilograms per hectare in 1962, has not yet doubled the level of the 1890's. Under a totalitarian regime, the Communist Chinese may accelerate the development of the chemical fertilizer industry faster than did the Japanese, or the Chinese in Taiwan. But resource constraints will impose obstacles that the Communist planners must find ways to overcome.

In the first place, the cost of building chemical fertilizer plants capable of producing 50 to 60 million tons a year would be prohibitive. It was estimated that, during the First FYP, the construction of a nitrogenous fertilizer plant with a capacity of 20,000 tons cost 100 million *yuan*, with a gestation period of four years (Wang and Han, 1957, pp. 11–15). To add 50 million tons to existing capacity would thus cost about 25 billion *yuan*, more than half the total state investment during the First FYP.

Second, there is the problem of obtaining adequate supplies of raw materials. A tremendous development of electric power would be necessary to supply the nitrogen needed. Other short materials include phosphate and

---

26. Speaking in January, 1956, Liu Lu-yen, then Minister of Agriculture, in explaining a draft of the National Program for Agricultural Development during 1956–67, envisaged a minimum requirement of 20 million tons of nitrogenous fertilizers per year. 1956, p. 38. When phosphorus and potassium were added, some 33 million tons of gross chemical fertilizers would be required annually. Dawson estimates that if chemical fertilizers were applied in areas where foodgrains and other crops could use them profitably at a rate equivalent to that currently used in Taiwan, some 50 to 60 million tons of gross chemical fertilizers would be required. Buck *et al.*, 1966, p. 114.

sulfuric acid.[27] The latter is widely used for other industrial products, and the chemical fertilizer industry would have to compete with them for the available supply. Measures have been taken to encourage the production of chemical fertilizers which do not require sulfuric acid (J. C. Liu, 1965, pp. 915–32).

Third, there are the difficult tasks of supplying machinery and equipment, mastering techniques of production, and training an adequate technical labor force. The Chinese have made considerable progress in imitating Soviet and Western methods of producing chemical fertilizers and in developing new techniques of their own. They are also able to design and build ammonia plants and supply these plants with domestically produced machinery and equipment. But the crucial factors here are the speed and efficiency with which new plants can be designed and constructed, and machinery and equipment installed in them. Moreover, the Chinese are still not able to produce chemical fertilizers that require relatively sophisticated methods. The new complete plants recently imported from the West may have demonstration effects, but the process of learning and imitating will be slow.

A fourth vital factor is the establishment of a transportation network to assemble raw materials for chemical fertilizer plants scattered around the country and to distribute the products to tens of millions of farms. The tonnage to be shipped would take up a large part of the country's transport capacity, and so would require a great deal of improvement and expansion of the existing transport system.

Finally, even if Chinese agriculture were adequately supplied with fertilizers there is the problem of their efficient utilization. A given type of chemical fertilizer cannot be applied universally to all soils. Soil needs will have to be determined scientifically before application. The experience of Japan indicates that use of chemical fertilizers requires careful coordination of seed, irrigation, soil preparation, and organic manures. It will take a long time for Chinese farmers accustomed to traditional methods of cultivation to learn new techniques. As the Chinese themselves admit, they are "still at the very beginning of a long process, and success can be expected only from a new, future, systematically educated generation of farmers" (*China News Analysis*, Feb. 15, 1963, p. 2).

Since 1963, the Chinese have once again placed emphasis on the development of small plants. During the Great Leap Forward, numerous small local plants were built to produce chemical fertilizers, and the results were disastrous. A new policy was adopted in 1962 "to establish a large number of new plants, especially large and medium-sized plants in a well-planned manner" (*Worker's Daily*, Dec. 8, 1962). But in 1963 small plants appeared again, and

---

27. The phosphate deposits in China as determined by the National Geological Survey in 1946 amounted to about 47,586,000 tons, which would last only a few years if China were to develop the chemical fertilizer industry rapidly. Recently, more extensive and accessible sources of lower grade phosphate have been found, but they are mostly located in remote regions and costs of transportation would be extremely high. Resources of pyrites, which produce sulfuric acid, are also limited.

it was officially stated in March, 1964, that "small plants are preferred because they require less investment and are easy to build and operate" (Quoted in MacDougall, July 1, 1965, pp. 14–16). It was pointed out, moreover, that transportation and storage problems would be eliminated, since fertilizers could be used directly by local farms.

The small plants built since 1963 were a great deal larger than the ones established during the Great Leap Forward, and were mainly mechanized.[28] Factories with annual capacities of 100,000 tons are working in a number of large cities, such as Shanghai, Canton, Kaifeng and Taiyuan. Small plants of a 2,000–5,000 ton capacity have been built in nearly every Chinese province. In Shangtung Province alone, 16 new chemical fertilizer plants started production in 1967 (*Far Eastern Trade and Development*, Oct. 1967, p. 965).

It would seem beyond the capabilities of the Chinese economy to support a program of fertilizer use substantially larger than the present one. The development of small plants may alleviate the difficulties to some extent, but is by no means an adequate solution. To increase agricultural production, measures other than fertilization are also necessary.

In the early days of the Communist regime, farm mechanization was regarded as the most important means of raising agricultural production. It was one of the great benefits which the government promised to bring to the peasants during the land reform of 1950–52. It also was considered by the leadership as a precondition for agricultural collectivization. But the policy was reversed in 1955 after a heated debate within the Party, which culminated in Mao's speech on agricultural cooperation in July of that year. Collectivization was given priority over mechanization.

Although mechanization was thus relegated to a status of secondary importance, considerable effort has been made during the last decade to expand the use of farm machinery and equipment. At first, mechanization was largely identified with the use of tractors. The emphasis was later placed on the adoption of mechanized irrigation and drainage equipment. Present policy stresses so-called "semi-mechanization," the introduction of new small farm implements and the improvement of existing ones.

In 1949, there were only 401 tractors in the entire country, mostly in Manchuria. The number rose to 59,000 in 1959 (expressed in standard units of 15 horsepower. N. R. Chen, 1967, Table 5.103). It has been claimed that there were more than 80,000 in 1960 (Cheng et al., 1966, pp. 454–461) and 100,000 in 1963 (*Wen-hui-pao*. Nov. 8, 1963). Western observers believe that, as of early 1966, China probably had 110,000 tractors, operating primarily in state fields. But even that number would have been capable of servicing only about 11 million hectares of cultivated land, less than 10 per cent of the

---

28. It was reported that the productive capacity of small local plants had been doubled every year since 1961. The output of small plants then accounted for only 2 per cent of total output of nitrogenous fertilizers, rose to 12.4 per cent in 1965, and was expected to reach over 18 per cent in 1966. *People's Daily,* June 16, 1966, p. 2.

total. Although great strides in the use of tractors have been made, the number in use is far below a "minimum" requirement of about 800,000 to 1,200,000.[29]

Several factors have hindered the development of the widespread use of tractors. The most obvious is lack of capital and technology, the chief obstacle to the development of almost every sector of the Chinese economy. Only one tractor plant, the Loyang First Tractor Factory, was built during the First FYP, largely with Soviet technical aid. Several other plants have since been added, in Tientsin, Nanchang, Changchun, Anshan, Shenyang, Wuhan and Hangchow, with a total annual capacity of 20,000 to 30,000 standard units. Since the withdrawal of Soviet experts Chinese technicians have been groping for new techniques of production. It was reported in January, 1964, that China had imported 1,200 tractors of different varieties for the purpose of research and experiment (NCNA broadcast, Jan. 3, 1964). The Chinese tractor industry is now mass-producing six types of tractors.[30]

A second problem is the shortage of materials. The basic material used in manufacturing tractors is steel, which is required for many other industrial and military products. In addition to steel, a wide variety of other metals are also required. The "East Is Red" tractor, for example, consists of more than 10,000 parts requiring 450 different kinds of metal. When the Loyang First Tractor Plant was put into operation in 1959, most of the metals needed were foreign products. Although the variety of metals produced domestically has since been increased, there are still many which China can produce only in small quantity, if at all. Moreover, the quality of the metals and parts manufactured domestically sometimes does not meet the required standard (*People's Daily*, Jan. 7, 1963, p. 1).

Apart from the factors detrimental to the production of tractors, there is lack of effective demand due to the peasants' low purchasing power. With the agricultural surplus largely extracted by the state, the amount of savings left in the hands of the peasants is small. The prices of domestically manufactured or imported tractors have been fixed so high that, apart for a few large, rich

---

29. According to the President of the Chinese Academy of Agricultural Mechanization, the minimum number of tractors required to plow the 107 million hectares of cultivated land would be 1,100,000. *Workers' Daily,* Jan. 7, 1963. Liu Jih-hsin's estimate indicates that at least 800,000 tractors would be needed to work the 80 million hectares of cultivated land which are believed suitable for the use of tractors. *People's Daily,* June 20, 1963.

30. These are as follows: the 100 h.p. "Red Flag" caterpillar tractor, manufactured for land reclamation by Anshan Tractor Plant, is adaptable for use as a bulldozer, grader, or mobile craner. The 75 h.p. "East Is Red", produced by the Loyang First Tractor Plant, is designed mainly for use in North China to perform such farm jobs as plowing and harrowing of land, sowing of seeds, harvesting of crops, and compacting of the soil. Another smaller "East Is Red" with 28 h.p. can be used in fields of maize and cotton for cultivation, fertilizer spreading and transport. The 35 h.p. "Bumper Harvest" tractor produced by the Hangchow Tractor Plant is an all-purpose model to be used in the paddy fields. The "Iron Ox" wheeled tractor produced by the Tientsin Tractor Plant can be used for plowing and harrowing land and for sowing seeds and pulling combine harvesters. Finally, the 7 h.p. "Worker-Peasant" is to be used in gardens, orchards, and terraced fields. Kuo, 1964, pp. 134–50; and *Far Eastern Economic Review, 1966 Yearbook.*

communes, only state farms can afford to employ them.[31] The use of a tractor frequently is not profitable when the increase in efficiency is not large enough to offset the high cost of the tractor. Chou En-lai once said that "the cost of production in many of the state farms was higher than the cost of production in cooperative farms where only human or animal power was used" (Government of India, 1956, p. 165).

The use of tractors is hampered further by a shortage of fuel. Although the supply of petroleum products has improved considerably, China is still short of petroleum, and military and industrial uses continue to receive priority. Lack of training in driving and maintenance also discourages farmers from acquiring tractors (Teng Chich, 1966, p. 5).

A number of natural and institutional factors serve further to handicap the introduction of tractors. Topographically, not all the farmland in China is suitable for tractor use. According to one estimate, approximately 65 per cent of the arable land can be cultivated with farm machinery (Chao Hsueh, 1957, pp. 16–18). But due to differences in topography, farm layouts, and methods of cultivation, a tractor cannot be used indiscriminately in different parts of the country and on all farms within the same region. A wide variety of tractors will thus have to be designed and manufactured.

Moreover, Chinese farms are still too small for effective use of tractors. On the basis of Soviet experience, the rate of utilization would reach 64 per cent of full time on a 33-*mou* farm and 85 per cent in a 1,500-*mou* farm (*Ibid.*). In 1962, some local Chinese authorities issued instructions that farms smaller than ten *mou* in size should not be cultivated with caterpillar tractors, and that wheeled tractors should not be used in farms of less than 5 *mou* (Kuo, 1964, pp. 134–50). The problem is that not too many farms in China average more than ten *mou*.

Within the small farm there are usually several fields. Dividing the fields are either irrigation canals, ditches and ponds, or trees. A large part of the three million mulberry trees in the country are planted in the boundaries (Chao Hsueh, 1957). There are also abandoned wells and ditches not yet flattened and boundary stones not yet removed. Some roads, bridges, and tunnels between the farms are not wide or strong enough for the passage of large tractors. To make tractors feasible, all these features will have to be adjusted. This will require a considerable amount of investment and no short period of time.

Last but not least, the relative abundance of rural labor and the extreme scarcity of capital makes the large-scale use of tractors uneconomical. Although tractors were hailed as one of the great innovations that Communism could bring to China, an idea inherited from the mystique of Soviet Communism, the Party realized that to attempt the use of tractors on a widespread scale

---

31. According to a survey of seven machine tractor stations in North China in 1958, a tractor and its attachments cost 15,000 *yuan,* which would have accounted for a large proportion of the annual savings of an average commune. Chin, 1965. Moreover, each tractor generally lasts ten years, during which about twenty major and minor repairs are required. The cost of replacing parts would be two to 2.5 times of the price of the tractor itself. *Workers' Daily,* Jan. 17, 1963.

would aggravate the already serious problem of unemployment and underemployment. By adopting an agricultural policy which stresses full utilization of idle labor, the Chinese apparently have abandoned the possibility of employing tractors on a significant scale in the foreseeable future.

Instead, considerable attention has been given recently to electrified and mechanized irrigation and drainage equipment for raising water to high lying fields and for draining waterlogged farmland. Lack of such equipment was considered partly responsible for the failures of the irrigation system during the crop crisis of 1959–61.

By the middle of 1965, 90 per cent of the *hsien* had some mechanized irrigation and drainage equipment designed and produced in China (*Far Eastern Economic Review, 1966 Yearbook*), while total capacity of mechanized pumping amounted to seven million horse power, eleven times the 1957 level (Elliot, 1965, pp. 50–51, and *Ta-kung-pao*, Oct, 1, 1962). In 1966 nearly 6.6 million hectares of farmland in China were irrigated with mechanized pumps. The efficiency of both irrigation and drainage equipment was claimed to have improved a good deal during recent years.[32]

The rationale behind the Chinese decision to expand the use of irrigation and drainage equipment seems to be based on land-saving and cost considerations. In view of labor redundancy in China, one might question the wisdom of substituting capital for labor even here. But the Chinese argue that the use of mechanized equipment for irrigation and drainage will save land which has become increasingly scarce in relation to labor, since traditional methods of irrigation require much more land than methods involving the use of mechanical power. To irrigate 100 hectares of land, 3.6 hectares are required for irrigation works using traditional, labor-intensive methods, but only 0.25 hectares if electric and mechanical equipment are employed (Wang, 1965, pp. 40–46). Moreover, irrigation by human and animal power may be too inefficient and slow to resist the sudden advent of heavy rains and floods.[33] It was suggested in the Chinese discussion that the labor saved through mechanized irrigation could be diverted to other farm tasks, such as more careful weeding, and to the subsidiary production which made such an important contribution to economic recovery after the Great Leap fiasco (Wang, 1965, pp. 40–46).

32. In 1965 mechanized pumps were serving some 80 per cent of the fields requiring irrigation on the Yangtse delta. *Far Eastern Economic Review, 1966 Yearbook*. As of June 1966, 43,000 units of mechanized pumps had been installed in the Southern provinces. For example, the distance which a pumping station could lift water rose from about ten meters in the early nineteen-sixties to 150 meters in 1966, and the acreage served by the stations had expanded a hundredfold during the previous several years. *People's Daily*, June 1, 1966, p. 2.

33. According to one estimate, one horsepower of mechanical equipment can irrigate about 3.3 hectares of land, equivalent to the use of three oxen. New China News Agency broadcast, Dec. 2, 1961. Another estimate indicated that to irrigate 667 hectares of paddy fields through electric power would save 2,000 human labor units. *Chinese News*, Nov. 18, 1961. A third estimate was that the cost of irrigation with the use of electric power was between one-sixth and one-fourth the cost of using human or animal power. *People's Daily*, Dec. 23, 1961, p. 1.

Chinese policymakers have come to the realization that it will take decades for China to modernize its agriculture. In October, 1964, the Central Committee of the Communist Party announced that "within a fairly long period the work of agricultural technical reform should be based on the policy of developing mechanization and semi-mechanization simultaneously, with the emphasis on semi-mechanization" (NCNA dispatch, Oct. 15, 1964). In January, 1966, the Vice Minister of the Second Ministry of Light Industry stated:

Only on the basis of carrying out gradually semi-mechanization of farm implements, can agricultural mechanization be gradually realized. It may be said that to develop semi-mechanized farm implements is the inevitable road toward agricultural mechanization in China (Teng Chich, 1966, p. 5).

In 1966 there were said to be some 25,000 handicraft workshops and producer cooperatives, with a labor force of more than 810,000 persons, specializing in the production of farm implements. More than 1.3 million pieces of improved and semi-mechanized implements of over 100 different varieties were produced by these workshops in nineteen Chinese provinces during the first nine months of 1965 (Chia, 1966, p. 5).

The use of semi-mechanized farm implements had been receiving a good deal of attention long before the 1964 policy announcement. But there was no appreciable increase in their use for several reasons. First, there was the acute shortage of steel and other materials necessary for manufacturing and repairing farm tools. The use of native iron and steel and other cheap and inferior materials, particularly during the Great Leap years, resulted in breakdown and deterioration of millions of implements (Kuo, 1964, pp. 134–50, and *People's Daily*, Jan. 4, 1966, p. 5). A survey of four communes in 1960 indicated that production could be raised by as much as 15 per cent if enough implements were added or repaired (*People's Daily*, Nov. 18, 1960, p. 3).

Second, there were defects in designing and planning farm equipment. In the 1950's, when a new or improved type of farm equipment was decided upon, the usual practice was to launch a campaign undertaken largely by village cadres among the peasants. Official statistics indicate that rapid progress in the popularization of certain types of implements was made during 1952–1955. But the upward trend did not continue for most of the implements, for the peasants began to resist purchase of those implements which did not perform satisfactorily. The outstanding examples were double-wheel, double-share and double-wheel, single-share ploughs. It was intended in 1956 to spread the use of some six million of these ploughs throughout the country within three to five years. They had never been properly tested under different soil and climatic conditions. Many of them were sold to the farmers indiscriminately and were later found not suitable for some areas, for example, areas where there was terrace cultivation, where there were waterlogged fields, or where the ploughs had to be drawn by buffaloes. In consequence, only 100,000 to 150,000 units were purchased by farmers in 1956 (Editor, *Economic Research*, 1956, pp. 1–4). State sales dwindled to 72,000 units in 1957, with 800,000 units left unsold (Kuo, 1964, pp. 134–50). The planners began to realize that popularization of equipment through propaganda might have undesirable consequences, and that adequate experimentation in different localities and for different crops was necessary before the recommendation of an implement to a farmer.

Finally, there was mismanagement of farm implements in collectives and communes. Before collectivization, farm tools were among the most important possessions of the Chinese farmer and were carefully maintained and repaired. As an editorial in the *People's Daily* put it, "a hoe would last three generations, and it was the property of the man who used it, repaired it, and cared for it" (*People's Daily*, Nov. 15, 1960, p. 1). With collectivization and communization, implements owned and used in common increased in number. With the loss of ownership, the farmer was no longer interested in taking good care of the implements. It was said that during the Great Leap years, "a great number of implements were lost, wasted, and destroyed"; and that "they were left scattered in the open air in the fields where rains and winds ruined them" (*Ibid.*). It was not until the latter part of 1960 that responsibility for farm tools and equipment was fixed for the commune and its production brigades, production teams and individual members (Kuo, 1964).

Current Agricultural Policy

The agricultural policy of China has been designed to be both extractive and developmental. In the first decade of the Communist government, organizational reform was employed as the most important means of executing this policy. It was successful in the extractive aspect, but failed on the developmental side. The agricultural growth rate contributed very little to the overall growth rate of the economy. With agricultural reorganization having reached its limit in extracting a surplus from the countryside, the Chinese have begun to take more effective measures to stimulate agricultural development. A long-term development policy was crystalized after the food crisis of 1959–61.

The policy calls for a gradual transformation of all farmland in China into "stable-and high-yield" fields. In the words of a New Year's Day editorial of the *People's Daily*, "the purpose would be to extend the area of farmland on which crops are protected and output is stable and high . . . to provide in a relatively short period a more reliable guarantee for the basic needs of the Chinese people in food and cotton" (Quoted in P. H. M. Jones, 1964, p. 350).

But the resources which the state and the nation as a whole can invest in the countryside are limited. Stable and high yields are to be achieved gradually in the course of two or three decades, basically by full utilization of underemployed rural labor. The state will invest its resources in certain key agricultural areas to push forward major types of capital construction, including "large-scale water conservancy projects, electrical engineering projects, soil improvement projects and forest shelter belts," (*People's Daily*, March 11, 1964), all of which require large capital investment and advanced technology. Small projects, and those portions of the major projects which can be done with labor-intensive methods, are to be carried out by the peasants using local resources, supplemented by government subsidies or loans.

Two categories of farm land are distinguished: ordinary fields and fields with stable and high yields. Fields in the former category include those with yields which are unstable but high, stable but low, or both unstable and low, (Lin Hung, 1964, pp. 12–21, 26), while the latter are those destined to produce "good harvests irrespective of drought and excessive rainfall and provide stable, high yields" (*People's Daily*, March 11, 1964). The state will concentrate its investment in stable-and high-yield fields which produce food grains, cotton, and oil-bearing crops with a high rate of marketable output (Lin Hung, 1964, and Hsu, 1965, pp. 15–19). By concentrating resources in key areas, the state may expect the highest return from its investment.

The communes are urged to mobilize idle labor for capital construction work, with the objective of stabilizing and then raising yields. Agricultural loans are granted on easier terms to those communes which have achieved favorable results. In November, 1963, the Agricultural Bank of China was established "to unify the administration of state investment and loans for agriculture."[34]

Sometimes the peasants are required to assist the state in completing large-scale key projects. But mostly they engage in such small projects for their own fields as building dams and storage pools, and digging ditches and irrigation canals to link up with the key projects. Most construction work is carried out during the winter months. Many communes also maintain special teams throughout the year for the purpose of maintaining and repairing construction projects; a commune usually allocates 20 per cent of its manpower to such work (Agricultural Economics Group, 1964, pp. 491–56, and 1965, pp. 1–13).

Not all idle labor in the rural areas is used for construction. Some of it is diverted to subsidiary production. Special efforts have been made to expand and diversify farm subsidiary activities in order to increase the sources of income for both the peasants and the state. The bulk of peasant savings from subsidiary income is used for the purchase of equipment and materials and for the payment of wages in capital construction projects (Chou, 1966, pp. 28–33).

The Chinese policy makers believe that with selective investment of state resources and efficient utilization of rural labor, modernization of agriculture will be achieved eventually, and that only then will agriculture be able to provide an adequate surplus for the continuing growth of industry. The

---

34. Most of the loans have gone to major grain and cotton producing areas. P. H. M. Jones, 1964, p. 350.

current policy appears to be a cautious and well balanced one likely to yield good results. In fact, it made significant contributions to recovery from the food crisis of 1959–61. Whether the Communist leaders will have the patience to continue the present program, which does not appear to have been greatly affected by the Cultural Revolution, and to forego new and violent experiments, remains to be seen.

CHAPTER FIVE

Population and Employment

If it were true that the wealth of a nation lay in its people, China would be the richest country in the world. Such a statement might be apt for a country which has reached an advanced level of development, but it is most decidedly untrue of contemporary China. The Communist regime has been engaged in a constant struggle to keep food output rising as rapidly as population. The liability of additional mouths outweighs the asset of additional hands.

Nevertheless, it is the same large population which in the long run must provide the sinews for growth. If the Chinese people, with their traditional dedication to hard work, could be deployed in an economically effective manner, the results might be astonishing. One has only to look at contemporary Taiwan for an idea of what could be achieved by a properly directed Chinese society.

In this chapter, we shall survey briefly the population and manpower resources of China. Where data exist at all, they are rudimentary, and we shall have to engage in a considerable amount of speculation. However, thanks to careful recent work by a few scholars, it is possible to suggest orders of magnitude with some claim to reality. The main lines can be accepted, though the details may not always be clear.

Population

There is no general agreement on the precise magnitude of the Chinese population. Various estimates for 1964 and 1965 are presented in Table V–1. The only one that is based on detailed demographic analysis is that of John

TABLE V–1
ESTIMATES OF THE POPULATION OF MAINLAND CHINA,
1964–1965

Source	Year	Estimated total population (millions)
John S. Aird	1965	715–875
Arthur G. Ashbrook	1965	760
Alexander Eckstein	1964	730
Chi-Ming Hou	1964	754
Edwin F. Jones	1965	728

Sources:
Aird, 1967, p. 363.
Ashbrook, 1967, p. 36.
Eckstein, 1966, p. 245.
Hou, 1968, p. 332.
Jones, 1967b, p. 81.

S. Aird of the U.S. Bureau of the Census, who offers a wide range of possibilities based on a variety of assumptions, rather than making an estimate in any strict sense.

Among the difficulties involved in determining the population of China is the fact that there are no satisfactory benchmarks. The only national census ever conducted was that of 1953, which yielded a total of 582.6 million people. It has been criticized as deficient on many counts by demographers, but Aird is of the belief that it can be used for estimation purposes:

With all its defects, which are undoubtedly many and serious, and despite the exaggerated official accuracy claims on one side and the sometimes immoderate scepticism of foreign analysts and commentators on the other, it must be conceded that the reported 1953 census total of 583 million is probably the nearest approach to a reliable population figure in the history of China. (1968, p. 244).

All the estimates in Table V–1 start with the 1953 census. Aird, however, does not accept the age-sex structure of the 1953 census, which appeared, among other things, to contain a serious undercount of females in the younger age groups, which is probably attributable to the traditionally lowly status of unmarried girls. Aird has gone back into Chinese demographic history to construct a model more consistent with modern ideas about population structure. This task was greatly complicated by the fact that China was plagued by a series of catastrophes resulting in large population losses. Between 1851 and 1953 there were no less than 16 major catastrophes—floods, famines, wars—which led to the loss of between 48 and 98 million lives. The first of these, the Taiping Rebellion, which ravaged China from 1850 to 1866, resulted in somewhere between 20 and 50 million deaths.

Even if one accepts the 1953 census as a starting point[1] and derives a more

---

1. Aird, in fact, feels that the 1953 census understated the population total by somewhere between 5 and 15 per cent, and some of his projections take this into account.

reasonable age-sex structure, there still remains the formidable problem of estimating subsequent growth rates. Neither politically nor economically has Communist China followed a "normal" path of development, in comparison with the experience of other nations. What, for example, were the effects of agricultural collectivization? Of the economic crisis after the failure of the Great Leap? Of the civil war that began in 1966?

Aird approaches the problem by constructing four alternative models for the period 1953–1965, each based upon a different set of birth and death rates drawn from varying assumptions about the impact of these catastrophic events. It is assumed, for example, that mortality rates fell until 1958, when further reduction could not have occurred without more substantial improvement in diet and medical care than China could afford. Various assumptions are made about the impact of the post-Leap crisis on mortality and fertility. Rates of natural increase under the alternative assumptions of the models range from a 1.75 per cent minimum per annum to a 2.25 per cent maximum. The width of the range of Aird's estimates in Table V–1 is attributable to differences in the net growth rates and in the degree of understatement assumed for the 1953 census.

The consensus among those who have studied this problem is that the true total for 1965 was much nearer the lower than the higher end of Aird's range. There is some evidence that the Chinese government accepted a population total of 680 to 690 million in mid–1963, though Mao personally felt that it was lower. (Aird, 1967, p. 351). It seems reasonable to conclude, as does Jones, that the population of China in 1965 was in the vicinity of 728 million, though the possibility of higher or lower figures cannot be ruled out.[2]

The full implications of these figures are staggering. If one combines the populations of the next most populous nations of the world, India, the Soviet Union, and the United States, the combined total is only about 150 million greater than that for China alone.[3] Moreover, the natural increase rates of these countries are probably lower on the average than that of China. In 1965, the three countries together added an approximate 17.5 million to their populations, while the increase for China was on the order of 16 million. *Each year*, China must feed, clothe, and house an *increment* to its existing population equal to the entire population of Scandinavia, or more than one-third more than that of Australia.

The conclusion is inescapable: rational policy for China involves a massive program of birth control. Until 1954, the Chinese adhered strictly to the Marxist idea that the Malthusian population trap did not apply to a Communist society. Mao came out squarely for this conception in the following terms:

---

2. Chinese official sources continued to cite a figure of 700 million in various proclamations issued in 1966, but it is difficult to reconcile so low a figure with what is known of the previous demographic experience.

3. Estimated 1965 populations for these countries are: India, 482 million; the Soviet Union, 231 million; the United States, 195 million; total, 908 million.

A large population in China is a blessing. We can manage our country well even if her population is further increased several fold. The solution lies in increasing production. The fallacy maintained by Western capitalist economists such as Malthus that the increase of food lags behind the increase of population has not only been theoretically refuted long ago by Marxists, it has been disproved by the facts existent after the revolution of the Soviet Union and in the liberated region of China.[4]

In 1954, perhaps as a result of the facts revealed by the census of the previous year, the Communist Party began to realize that whatever the long run prospects, population might well outstrip production in the short run, and a drive to reduce births was begun. Contraception, abortion, and sterilization were encouraged. Hardly had the program gotten under way when the enthusiasm of the Great Leap Forward, with its emphasis upon mass labor deployment, brought it to a halt.

The post-1960 depression, bringing in its train severe food shortages, gave rise once more to the conviction that birth limitation was a practical necessity, whatever ideology might dictate. Considerable progress has been indicated for some of the larger cities since 1963. Strong economic and social pressures have been brought to bear on young people to prevent marriage below prescribed ages. Birth control clinics have been established, and contraception made readily available. There appears to have been a considerable decline in the birth rate where these programs were undertaken, but thus far the countryside, where most Chinese live, seems unaffected.

The future course of Chinese population depends upon a number of variables above and beyond the vagaries of official policy. Normally, in the course of economic development, population is rapidly boosted in the early stages of growth by a combination of falling mortality rates and persistently high fertility rates. As per capita income rises, birth rates begin to fall; families have an increased inclination and ability to limit the number of children.

A crucial question for China is the rapidity with which per capita income can be expected to rise. This depends in part on the course of agricultural production and the portion of the national product which is devoted to consumption. The effectiveness of the distribution system, the prevalence of law and order, and the degree to which the traditional desire for children on religious grounds can be overcome will also have an impact on population growth, upon mortality as well as fertility.

It is extremely difficult to specify these variables with any degree of precision and to relate them functionally to population change. Increasing per capita income may, under certain circumstances, lead to an increase rather than a decline in births. Social turmoil may lead to an increase in the death rate, but at the same time may give rise to a higher birth rate by interfering

---

4. Quoted by Edwin F. Jones, 1966, p. 10. We are indebted to this source for the paragraphs on birth control.

with birth control campaigns. Arguing by analogy from the experience of other countries is a hazardous procedure in the Chinese case.

Aird explores the problem by setting up a number of models based upon alternative hypotheses of future events, ranging from "maximum development," which assumes that the Chinese leaders have learned from the mistakes of the past and will follow a steady and prudent course, to "catastrophe," defined as the collapse of social order, with major famines beginning in 1970 and continuing for several years. The first yields a total of 1.3 billion people in 1986, the second, 925 million. The Chinese themselves have indicated that they anticipate a population of around 1.05 billion in 1986 (E. F. Jones, 1966, p. 23).

Even under the most moderate view of China's population growth, therefore, during the two decades 1966–1986, the *addition* to the population of China will be almost equal to *total* 1966 United States population. If the Chinese government is correct, a new population equivalent in size to the combined present populations of the Soviet Union and Japan will have been created. And if the "optimistic" Aird projection should prove true, another India will have been added to the world. It is difficult to exaggerate the Chinese population problem.

Nonagricultural Employment

The very concept of employment in underdeveloped economies is a difficult one. Usually, a large proportion of the population is engaged in agricultural activity which is either of a subsistence character or, if producing for the market, of low productivity. Much of the urban population is occupied with marginal services which are sometimes difficult to distinguish from unemployment. The modern sector, though critical for development, is typically quite small.

China is clearly a dualistic economy in this sense. There is a modern urban employment sector for which data are available for the 1950's. Statistics for the traditional handicraft and service sectors are rudimentary, and the extent of this activity can only be guessed at. Rural employment data are non-existent, and one can operate there only within very broad margins of error.

Starting with what is best known, we present in Table V–2 data on industrial employment for 1952 and 1957, the first and last years of the first Chinese Five Year Plan. Despite an increase of 43 per cent during the period, employment in this sector remained minute in relation to population in 1957. The tremendous industrialization drive netted the country only two million more jobs in modern industry.

The greatest gainer within manufacturing was heavy industry. The metals, chemical, and oil industries all underwent a considerable expansion of employment; consumer goods advanced much more moderately. But even in heavy industry, the absolute increases were not great. For example, pig

TABLE V–2

EMPLOYMENT IN CHINESE INDUSTRY, 1952–1957
(thousands of persons)

Industry	1952	1957	Percent increase 1952–1957
Electric power	64	143	123
Coal	494	669	35
Petroleum	22	67	205
Iron and steel	233	347	49
Nonferrous metals	158	346	119
Metal processing	846	1403	66
Chemical processing	113	253	124
Building materials	421	600	43
Paper	84	94	12
Textiles	1022	1282	25
Food processing	1021	1200	18
Total	4478	6404	43

Source:
Emerson, 1965, p. 143.

iron output increased from 1.9 to 5.9 million tons during the period, but because of rising productivity, there were only 114,000 more men employed. Development of the modern manufacturing sector, as the Chinese did it, was obviously not a path to full employment.

The data on nonagricultural employment in Table V–3 take us to the next step of aggregation. Employment increased by less than three million from 1952 to 1957, despite the fact that the urban population as a whole rose by 20 million (Hou, 1968, p. 342). The additional employment afforded by the modern sector was counterbalanced by reduced employment in the traditional sectors, with the handicrafts remaining constant. As in the case of manufacturing, productivity increases in the modern sector precluded employment increases in proportion to output.

The structure of nonagricultural employment in China is compared with that of the Soviet Union in 1928 in Table V–4. On the eve of the first Soviet Five Year Plan, a greater proportion of the Soviet labor force was already engaged in industry than in China after five years of intensive industrialization. Noteworthy, too, is the much higher relative commercial employment in China, reflecting the petty trade characteristic of underdeveloped countries.

It is difficult to evaluate Chinese employment statistics for the period of the Great Leap. The official totals for nonagricultural employment are:[5]

---

5. N. R. Chen, 1967, Table 11.1. These figures include only so-called "workers and employees," and would exclude most of those engaged in the traditional sector, and many of those in the services, who appear as part of the non-agricultural labor force in Table V–3.

1957—24,506,000
1958—45,323,000
1959—44,156,000

Of the huge increase in 1958, Emerson says:

> The absolute increase in that one year was about six times as large as the increase during the whole of the preceding five-year period... Even as large as it is, the increase... probably understates to a considerable extent the manpower added during the year. It does not include persons engaged for short periods without pay in mass labor projects. (1965, p. 78).

What is questionable, if not the total, is the economic meaningfulness of the increased employment. The doubling of the labor force within a year must surely have meant a marginal product near zero for a good proportion of the new recruits, who were presumably raw and untrained and could only be used for simple unskilled work. As we have already pointed out, many of the Great Leap projects were uneconomic in character, and some were actually of negative value.

Between 1961 and 1963 there was a reduction in urban population that may have been as great as 20 million persons (E. F. Jones, 1966, p. 86). In

TABLE V-3

NONAGRICULTURAL EMPLOYMENT IN CHINA, 1952 AND 1957
(thousands of persons)

Sector	1952	1957	Percent increase 1952–1957
Factories and mining	3,283[a]	5,907[a]	80
Urban utilities	41	133	224
Handicrafts	9,344	10,000	7
Salt extraction	270	470	74
Construction	1,048	1,910	82
Water conservancy	134	340	154
Fishing	1,336	1,500	12
Modern transportation	1,129	1,878	66
Traditional transportation	5,362	4,826	−10
Modern trade and catering	2,724	5,245	93
Traditional trade and catering	11,416	5,374	−53
Finance, banking, insurance	576	621	8
Personal and professional service	3,230	3,339	3
Government	4,971	6,108	23
Total	44,864	47,651	6

Note:
[a] These figures differ from the totals in Table V-2 by virtue of the fact that the latter include employment in factory handicrafts, fishing, and salt extraction, which are shown separately here.

Source:
Hou, 1968, p. 366–67.

TABLE V-4
THE STRUCTURE OF NONAGRICULTURAL EMPLOYMENT IN CHINA AND THE SOVIET UNION
(per cent of total)

Sector	China 1952	China 1957	Soviet Union 1928
Industry	32	38	45
Construction	2	4	8
Transportation	14	14	12
Commerce	33	24	9
Government and other services	19	20	26
	100	100	100

Source:
Hou, 1968, p. 368.

all likelihood, many of those who were employed for the first time in 1958 were among the returnees. Unfortunately, there are no data against which to check this hypothesis. It is also probable that the nonagricultural labor force remained permanently above the 1957 level, but highly unlikely that it remained at the 1958–1959 level.[6] These are wide limits, but until we learn something about the course of output and productivity in recent years, it will be difficult to be more precise. One may only guess that employment in the modern sectors has risen since 1957, and that the traditional sector continues to furnish about the same levels of employment as in 1957, which would mean an increase in total non-agricultural employment. With the population growing, many may have been forced into marginal urban trades, adding to the existing employment problem.

Employment in Agriculture

The concept of agricultural employment is less precise than that for industry, and statistics are non-existent. Farm labor requirements are subject to wide seasonal swings, affecting not only the number of family participants, but the intensity of work as well. Many rural inhabitants, moreover, are part-time farmers and part-time handicraftsmen.

It has been estimated that the agricultural population of China in the age group 14 to 64 was between 281.9 and 291.8 million in 1957, and in the age group 15 to 59, some 260 to 269 million persons (Hou, 1968, pp. 365–66). Most of these people were employed directly in farm work, though a portion either were performing domestic tasks or were in handicrafts or trade. Liu and Yeh derive a rough estimate of 5 per cent for the latter groups, based upon

---

6. Jones believes that industrial employment was roughly halved from 1960 to 1963, but there is no concrete evidence to support this belief. *Ibid.*, p. 86.

pre-war surveys of the pattern of rural work.[7] Applying this figure to the totals shown above, the following figures may be derived for the 1957 agricultural labor force: 14–64 years, 268 to 277 million; 15–59 years, 247 to 256 million.

These figures cannot be compared directly with the nonagricultural employment data shown in Table V–3. These purport to show total labor force, not merely employment; they include people working part-time, the unemployed, and the ill. But they do convey some sense of the overwhelmingly agricultural character of the Chinese economy. Somewhere on the order of 80 per cent of the Chinese people of working age were still engaged in agricultural pursuits in 1957. When this is compared with a range of 10 to 25 per cent for the developed nations of the world (and even less for a few) one has some sense of China's economic backwardness.

From 1950 to 1960, the urban proportion of the Chinese population rose from 11 to 19 per cent, on the basis of one estimate (Hou, 1968, Table 5). But this trend was reversed in 1961, when millions of people returned to the country. The switch of emphasis from industrial to agricultural investment, and the much lower rate of post-Leap industrial expansion, must have meant that most new jobs were in agriculture, if anywhere. Whether the Chinese have been able to arrest the cityward drift of population characteristic of underdeveloped nations cannot be known with certainty. But it would appear that the present proportion of the labor force in agriculture is at least as great as it was in 1957, and perhaps even larger.

The absolute size of the labor force has grown with the population. If we make the simple (and not very realistic) assumption that the agricultural labor force was increasing *pari passu* with the total population, the 1965 level might be 310 to 320 million (16–64 years) or 286 to 296 million (15–59 years). However imprecise these figures are, they do seem to point to an agricultural labor force of about 300 million in 1965.

By way of comparison, about 43 per cent of the 439 million people of India within the 15–65 year age group in 1961 were reported to be economically active, and of these, 69.5 per cent were engaged in agricultural pursuits (ILO, 1964, pp. 8 and 36). Applied to the Chinese total of 750 million, this would yield an economically active population of 322.5 million and an agricultural labor force of 224 million for China. This is substantially below our estimates; the data suggest that the labor force participation rate was considerably higher in China than in India. On the other hand, the Soviet participation rate of 52.2 per cent for 1959 would yield an economically active population of 392 million if applied to China, and this seems to be beyond any possible upper limit. Demographic and economic structure, as well as custom and tradition, determine the level of economic activity of a given population. Without going further into the matter, it seems fair to

---

7. Liu and Yeh, 1965, p. 102. The Liu-Yeh estimate is based upon rural population in the age groups seven to 64 years and may not be appropriate to the age groups specified in the text, but there is no other estimate.

say that the percentage of the Chinese labor force which is economically active seems to lie somewhere between India's 43 per cent and Russia's 52 per cent.

Unemployment

The Communist regime claimed that there were four million persons unemployed when it assumed power in 1949, and conceded that some unemployment prevailed for a while thereafter. But as in the case of other Communist countries, national unemployment statistics were not published. However, a survey of the province of Anhwei did suggest an unemployment rate of 9 to 11 per cent among urban male workers (Hou, 1968, pp. 372–73).

Estimates of rural unemployment and underemployment are so crude as to be of little value. Urban unemployment is a more tractable concept, both conceptually and statistically. Several estimates are shown in Table V-5.

TABLE V-5

ESTIMATES OF NONAGRICULTURAL UNEMPLOYMENT AMONG URBAN MALES IN CHINA, 1949–1960
(per cent of urban labor force)

Year	Ages 14–16 years[a]		Ages 15–59 years[a]		Ages 12–64 years[b] (Male nonagricultural, rural and urban)
	I	II	I	II	
1949	31.7	24.2	27.4	18.2	—
1950	27.6	19.8	23.1	13.2	—
1951	21.8	13.6	17.0	6.8	—
1952	22.3	14.3	18.0	7.5	34
1953	21.3	13.4	17.4	6.9	36
1954	26.2	18.9	22.7	12.7	38
1955	29.6	22.7	26.3	16.8	39
1956	29.2	22.4	26.3	16.5	38
1957	31.6	24.8	29.0	19.5	38
1958	14.9	7.1	12.5	0.3	—
1959	24.2	17.6	22.3	12.4	—
1960	28.3	22.3	26.3	17.3	—

Sources:
[a] Hou, 1968, p. 369.
[b] Liu and Yeh, 1965, p. 102.

Estimates I and II differ only with respect to the assumed proportion of the nonagricultural population—25.5 per cent for I and 23.0 per cent for II. The Liu-Yeh estimate is for nonagricultural workers in rural as well as urban areas.

How reasonable are these estimates? The Liu-Yeh figures seem impossibly high from what we know of unemployment in underdeveloped countries.[8] It is even difficult to credit some of the Hou estimates. The International Labour Office has observed that

> ... widespread open unemployment is not typically found in underdeveloped countries. The reason is simple. In the absence of unemployment compensation people without proper jobs cannot afford to be wholly idle. They will join relatives, lending a hand in the work to be done for example on the family farm, or they will employ themselves in some minor activity such as peddling, shining shoes, or begging. In so doing they do not displace other workers but cause the work to be spread more thinly over more people.[9]

The lower of the Hou estimates approaches the bounds of credibility. Whatever the absolute level, however, the trend shown in Table V–5 does find support in what we know of Chinese development. The capital-intensive new industry installed from 1952 to 1958 provided few new jobs. At the same time, the population was growing rapidly, and the cities even faster.

The sharp decline in unemployment in 1958 was attributable to the mass labor projects undertaken during the Great Leap Forward. It is somewhat difficult to understand the rise in 1959, but by 1960 retreat from Great Leap policies was in full swing, and there is little doubt that unemployment rose.

The ensuing decline in capital construction, the closing of manufacturing enterprises for want of raw materials, and the contraction of economic activity in general, may well have pushed unemployment up to unprecedented heights. The enforced movement of people out of cities was undoubtedly designed to alleviate the pressure of unemployment, and the gradual upswing since 1963 must have had the same effect. The Red Guard movement of 1966–67 suggests that there were still many idle hands in addition to the students who seemed to constitute the bulk of the irregular forces mobilized in the internal political struggle.

Professional and Scientific Manpower

An essential ingredient of economic development is technically trained manpower. Communist China was grossly deficient in this factor on the eve of its First Five Year Plan. In 1949, over 80 per cent of the population was illiterate. It was necessary not only to educate an elite capable of directing the economy, but to widen the entire base on which that elite was based.

Soviet assistance was of considerable importance to China in the first decade of Communist power. About 11,000 Soviet scientists and technicians

---

8. It should be noted, however, that the Liu-Yeh estimate represents the difference between population and employment, and is thus a rough estimate of full-time unemployment. Most of it presumably reflects underemployment rather than outright unemployment.

9. ILO, 1964, p. 124. Surveys of unemployment have yielded the following for other underdeveloped countries: Ceylon, 10–13 per cent (1959–60); Phillipines, 7 per cent (1958); Pakistan, 6–10 per cent (large towns, 1955); Indonesia, 7 per cent (urban, 1958); Chile, 6–8 per cent. *Ibid.*, pp. 23–24.

were sent to work in China, half in industrial enterprises, 18 per cent in transport and communications, and the rest in agriculture, education, public health, and scientific research. Over 700 Russians lectured in universities and technical schools, helping to establish new departments and laboratories.

Of no less importance was the education of Chinese in the Soviet Union. Some 38,000 Chinese received training in the Soviet Union between 1950 and 1960; half of them were workers, and the others included 7,500 students (of whom 2,000 were graduate students) and 1,300 scientists. Even after the cooling of relations in 1960, many Chinese continued their studies in Russia, and the last contingent did not leave until 1966.

Following is an appraisal of the qualitative impact of Soviet assistance in the sphere of training:

Almost every important branch of the technical or natural sciences in China was created or expanded with Soviet assistance during 1950–60 . . . The Soviet laid a foundation for the development of branches of science previously unknown in China. Departments formerly short of qualified personnel have gradually formed armies of experts, while those branches of science previously underdeveloped have advanced remarkably. (C. Y. Cheng, 1965, p. 206).

No one has yet been able to devise precise methods of measuring the contribution to economic development of the input of highly qualified manpower, but there is a growing consensus that it is a critical factor. It is not only a question of the immediate techniques and equipment that can be used, but of a stream of future benefits. Chinese nuclear physics, mathematics, geology, biology, medicine, and engineering were profoundly influenced by a decade of close contact with the Soviet Union. By way of comparison, from 1850 to 1962, some 24,300 Chinese received training in the West, and 12,000 in Japan. Of the total, 15,500 were scientists and engineers, of whom 4,500 were still alive and working on the Chinese Mainland in 1962. The largest group studied in the United States. A number of foreign-supported educational institutions functioned in China prior to 1949, providing additional Western impact.

The break with Russia caused China once again to look elsewhere for technical assistance. Japanese and European scientists and technicians have been finding their way to China in increasing numbers, and Chinese students have spread out in the West (e.g., 102 were sent to France in 1964, but left in 1967, as did most Chinese students abroad). Western influence remains strong among the older generation of Chinese intellectuals, and, until the events of 1966–1967, it began to look as though it might be strengthened. Soviet influence seems to be paramount in applied science and technology; in the basic sciences, the matter is more complicated. The Western and Japanese-trained scientists "constitute the backbone of Chinese professional, scientific, and technical manpower. They not only form much of the core of scientific personnel in the Chinese Academy of Sciences and other research organizations, but also they have become a leading force in institutions of higher learning" (C. Y. Cheng, 1965, p. 235).

Foreign assistance, critical though it may be in providing key personnel, cannot provide an adequate substitute for domestic education and training in the long run. The Communists greatly expanded the scope of their educational system, and, though quality may not always have been maintained, the quantitative results are impressive. Primary school enrollment, according to Chinese claims, rose from 29 million in the 1950–51 school year to 64 million in 1957–58; secondary school enrollment for the same periods rose from 1.3 million to 6.3 million; and higher school enrollment rose from 137 thousand to 660 thousand (Orleans, 1961, pp. 33, 35, and 61). Huge increases were claimed for the Great Leap period, but these must be discounted.

Again according to the Chinese, 50.7 per cent of all children of primary school age were attending school in 1953; 61.3 per cent in 1956; 85 per cent in 1957 and 1958; and 87 per cent in 1959 and 1960 (*Ibid.*, p. 32). These figures, if they can be credited, suggest that China is approaching universal primary education. Secondary school education still has a long way to go, however.

In higher education, engineering absorbed from 30 to 40 per cent of all students between 1950 and 1960. The next largest groups were those undergoing teacher training (10 to 20 per cent), with physical science third. The number of law graduates was reduced to a trickle, though in pre-Communist China it was the largest single group.

The Communist emphasis makes considerable sense from the point of view of development. Too many lawyers and too few engineers are being trained in Latin American and African universities. It is undoubtedly true that quality is sharply down, if only because of shorter terms and inordinate amounts of time spent on political indoctrination. There are still specialties in which advanced degrees cannot be given because of the lack of qualified teachers. But when all is said and done, Communist China graduated 431,000 students from higher educational institutions from 1949 to 1958, against 211,000 in the 36 years from 1912 to 1947. Average quality was probably higher under the old regime. "Contributory to this was the more limited enrollment. Most of the teachers in the institutions of higher learning had received their advanced training in Western countries. Their academic backgrounds were far superior to those younger teachers who now make up the large majority of the faculties in the institutions of higher learning. One important factor contributory to the higher quality of the older scientists and engineers was that their studies were not interrupted by compulsory extracurricular activities" (C. Y. Cheng, 1965, p. 102). Nevertheless, a trade-off of quantity for quality may be justified in the context of development requirements.

Apart from formal academic institutions, there was a proliferation of "spare time" schools and in-plant training courses to upgrade skills. About 470,000 workers were enrolled in the former at the end of 1960; they studied evenings and weekends in courses that ranged in duration from two to five years.

Secondary vocational schools were also established, some attached to industrial enterprises, others independent. About 740,000 persons attended these institutions from 1949 to 1958, not a large number considering labor force growth during the period. Most workers must have acquired their skills on the job.

The Communists, at an early stage of their rule, embarked upon a widespread literacy campaign, concentrating particularly on industrial workers. Some startling claims were issued about the success of the campaign, but there is no way of assessing them. One problem is that literacy standards are elastic because of the nature of the Chinese language. Primary school graduates are expected to have mastered 4,000 characters, but half, or even less, of this number has been established as satisfactory in adult training.

It is estimated that in 1967 there were slightly more than 1.7 million graduates of higher educational institutions in China, one-third in engineering and one-third in teaching. Only 6 per cent of the graduates were in natural science, which may hamper the aspirations of the regime (Orleans, 1967, p. 510). "Many of the Chinese engineers . . . are overly specialized, extremely weak on basic theory, and, in effect, little more than middle level technicians. But China needs this type of specialist. It would be a waste of time and money for China to produce nothing but highly trained engineers who have spent four, five, or more years at a university. While a nucleus of these people is available, the Chinese economy, in the present stage of its development, would probably have difficulty in absorbing large numbers of these individuals" (*Ibid.*, pp. 513–14).

China has been cited as an exception to the rule that underdeveloped nations find it very difficult to install viable programs of skilled manpower training (Harbison and Myers, 1964, pp. 88–91). An idea of China's situation can be garnered from a comparison with other countries at low levels of economic development. The relevant data are shown in Table V–6. They are rough approximations; the underlying classification of occupations varies from one country to another. But the data do suggest that in 1957, China was somewhat less well equipped with professional and technical personnel than India in 1961, substantially below Egypt, and considerably better off than Pakistan when measured against total population. In relation to the economically active male population,[10] the China-India-Pakistan ratios remain stable, but Egypt and Thailand rise in the scale. The reasons for the shift, which have to do with age-sex ratios and the rate at which men of working age participate in the labor force, need not detain us.

From 1957 to 1962, there was a very substantial increase in the number of professional and technical workers in China; the number of engineers and technicians rose from 496,000 to 1.4 million during this period, and the

---

10. The male, rather than the total, economically active labor force is used because of the extreme difficulty of determining what proportion of the females fall within this category.

TABLE V-6

PROFESSIONAL, TECHNICAL, AND RELATED WORKERS AS A PROPORTION OF TOTAL POPULATION AND OF THE ECONOMICALLY ACTIVE MALE NONAGRICULTURAL POPULATION, CHINA AND SELECTED COUNTRIES

Country	(1) Professional etc., workers (thousands)	(2) Total population (thousands)	(3) Ratio of (1) to (2)	(4) Economically active males, nonagricultural (thousands)	(5) Ratio of (1) to (4)
China (1957)	4,418[a]	637,000[b]	0.69	58,000[b]	7.6
India (1961)	3,238	439,235	0.74	40,535	8.0
Pakistan (1961)	414	90,283	0.46	7,003	5.9
Egypt (1960)	214	25,841	0.83	2,964	7.2
Thailand (1960)	174	26,258	0.66	1,570	11.1

Notes:

[a] This includes the following categories:

Engineering and technical personnel	496,000
Doctors and pharmacists	74,000
Feldshers	136,000
Nurses	128,000
Cultural and art personnel (1955)	88,500
Teachers	2,780,000
Midwives	715,000
Total	4,417,500

[b] The population total is from Liu and Yeh, 1965, p. 102. If the 1953 census data are accepted, the various estimates for this year are quite close. The figure for the economically active labor force is from Hou, 1968, p. 351 and covers the age group 14–16 years. The ILO age groups are 15 to 65, so that the Chinese total is subject to relative overstatement. No correction has been made, however, because of the approximate character of the unadjusted data.

The employees covered by the ILO data for professionals, etc., embrace Major Group O of the International Standard Classification of Occupations. The Chinese figures do not cover the clergy or jurists, but are otherwise consistent with the ISCO grouping.

Sources:
For China, N. R. Chen, 1967, Tables 11.1, 11.3, and 11.10; and Emerson, 1965, p. 93.
For other countries, ILO, 1964, Table 48.

number of doctors to 100,000, compared with 74,000 doctors *and* pharmacists in 1957. If one were to assume on the basis of these component groups that the total of the professional and technical groups doubled, the ratio of these groups to total population would have been 1.21 in China in 1962, far greater than that for any of the other countries in Table V–6. More conservative assumptions would scale the ratio down, but the conclusion seems justified that by the beginning of the 1960's, China was relatively well off in qualified manpower.

The question may be raised whether the output of Chinese educational institutions may not have been greater than the demand, with the result that many high school and college graduates could not be absorbed into the economy at their full skills. Student participation in the Red Guard units may have reflected some dissatisfaction with the employment opportunities facing them on graduation. The closing of all schools during the 1966–1967 academic year had the effect of easing the pressure on the demand for skilled jobs, as has the trend toward simultaneous admixture of work and education. But the problem is still there; hundreds of thousands of young men and women in the cities find the way up the social ladder blocked. Despite all the efforts of the propaganda machine,[11] China may discover that an underemployed intellectual class can menace the stability even of the Communist regime.

---

11. Orleans quotes the following advice to students concerned about their inability to secure admission to universities: "Provided we have a red heart and a wish to serve the people wholeheartedly and to labor, study, and struggle for the sake of the revolution, then we shall be able to make useful contributions to the motherland whether we continue our studies or take part in productive labor." (1967, p. 516).

CHAPTER SIX

The Control and Allocation of Resources

The economic goal of the Chinese Communist Party has been to build China as rapidly as possible into a great industrial and military power with a high degree of autarky. The main lines are determined by the ideology of the Party, the political ambition of the leadership, and the underdeveloped nature of the economy. The fundamental aim has remained unchanged during all the years of the Communist regime, although the strategies employed have varied.

To carry out the purposes of the Party required, first, the establishment of a system of state or collective ownership of the means of production and, second, centralized control of major economic activities. The elimination of private ownership of the means of production, which was itself a part of the national goal, was an important step in facilitating the implementation of central economic controls. Financing rapid economic growth required a high rate of capital accumulation which could not be obtained by voluntary saving. During 1931–36, a period of relative peace and prosperity in China, the rates of investment (the ratios of gross capital formation to gross domestic product) varied from 6.6 to 9 per cent per year (Yeh, 1968, p. 511). These rates were substantially lower than those achieved under the Communists. One way to compensate for a low rate of domestic saving would have been the importation of foreign capital. But the ideology of the Communist Party and its desire for economic independence made a large inflow of foreign capital unacceptable. Consequently, centralized measures for the mobilization of the savings potential were imperative.

Mobilization of savings, however, does not automatically lead to a high rate of capital formation, since the latter requires, among other things, stepping up the output of producer goods. A system of central planning had to be instituted to channel the bulk of the nations' investment resources into producer goods industries and supporting sectors, and to allocate the remainder among other sectors of the economy.[1] Relative neglect of investment in these sectors caused wages to outstrip the supply of food and other consumer goods. Therefore, state planning in the form of rationing of consumer goods and special taxes were also required to bring demand back into line with supply.

The underdeveloped state of the Chinese economy and lack of experience on the part of the planners made economic planning a difficult task. In the early years of the Peking regime, Soviet planning techniques were applied rather indiscriminately. These techniques were in many cases not suitable to Chinese conditions. There were many small production units both in the industrial and agricultural sectors, most of them employing primitive techniques of production. There were some six hundred million consumers among whom goods had to be rationed. These were formidable obstacles to central planning. Accordingly, markets, a price mechanism, and other decentralized measures were employed to back up centralized control.

The planning and management of the Chinese economy can be discussed under four heads: first, reorganization of the economy with the aim of nationalization or collectivization of the means of production; second mobilization of savings to finance industrial development; third, formulation and implementation of national economic planning; and finally, the use of markets and prices in resource allocation.

Economic Reorganization

When the Communists gained power in 1949, one of their first steps was to take over the major undertakings that had been operated by the Nationalist government. Large factories, railway and highway facilities, airlines, and a number of banking and commercial institutions, were all nationalized by the new regime; they became the backbone of its state enterprise system. In industry alone, the Communists seized nearly 3,000 enterprises, with total assets of from 10 to 20 billion dollars (C. Y. Cheng, 1963, p. 61). According to Communist statistics, state enterprises in 1949 already accounted for 41 per cent of the gross value product of modern industry, 68 per cent of coal output, 92 per cent of pig iron output, 97 per cent of steel output, 68 per cent of cement output, and 53 per cent of cotton-yarn output. At the start, all technical and managerial personnel were retained so the enterprises could continue to operate. But as soon as the new government had consolidated its control, a so-called "democratic reform" was launched to expel former management officials. Leading personnel in state enterprises were gradually replaced by party cadres and other politically reliable managers. The reform

---

1. Although agricultural failures in the early 1960's forced the Peking regime to transfer resources from such producer goods industries as machine-building to industries that would directly benefit agriculture, notably the chemical fertilizer industry, this change did not indicate a de-emphasis of producer goods production so much as a shift in emphasis from one set of producer goods to another. The bulk of state investment was still devoted to producer goods.

was reported to have removed several hundred thousand persons from state enterprises.

At the same time, the new regime began to impose restrictions on the activities of foreign enterprises. They became so strict after the outbreak of the Korean War that, by the end of 1952, most foreign firms found it impossible to continue operating and elected to leave China. Part of the foreign assets were transferred to the government without compensation.

The Communist practice was to absorb private enterprises into the state sector, not by outright nationalization, but through a process of gradual penetration. The first step was to ensure the continued operation of private factories by placing state orders with them for processing and manufacturing products for which raw materials were supplied by the state. Private trading units were allowed to become retail distributors or commission agents for state commercial organizations. All private enterprises, both in industry and in trade, were required to make full reports to the state on current transactions, their financial status, and other aspects of their activities, to submit production and sales plans to the state for approval, and to distribute their earnings between dividends, welfare funds, and other uses in proportions specified by government regulations.

Private enterprises developed rapidly during the Korean War, due to a substantial increase in state orders.[2] In order to check the growth of the private sector and to wipe out its profits, a drive known as the "Five Antis Campaign" was conducted during January-April, 1952.[3] As a result, a large amount of private capital was confiscated by the government. In industry, increasingly larger proportions of private facilities were required to produce for the state.[4] In commerce, private retailers were gradually replaced by cooperative stores and supply and marketing cooperatives, and by 1954 practically all the functions of private wholesalers had been taken over by state commercial organizations.

The next step taken by the government was to transform private enterprises to joint state-private operation, in which the government made an investment and appointed personnel to share managerial responsibilities with private owners. This policy had already begun in the earliest days of the regime, but at that time the process of assimilation of private enterprises was slow and cautious. The number of joint enterprises showed some increase after the five-antis campaign,[5] but developed more rapidly toward the end of 1955

2. The total value of private industrial output increased by approximately 70 per cent between October, 1949 and September, 1952.

3. The "five evils" of private enterprise fought against in the campaign were—bribery of government workers, tax evasion, theft of state property, cheating on government contracts, and stealing economic information from the government.

4. In 1949, the proportion of the gross value of products processed and manufactured by private factories for the state in the gross value of total private industrial output was 12 per cent. The proportion rose to 29 per cent in 1950, 43 per cent in 1951, 56 per cent in 1952, 62 per cent in 1953, 79 per cent in 1954, and 82 per cent in 1955.

5. In 1949, there were only 193 state-private enterprises, employing about 100,000 persons. By the end of 1953, the number of such enterprises had risen to more than 1,000, with an employment of 270,000 persons.

and the early part of 1956, coinciding with the "socialist upsurge" in the agricultural sector. At the end of 1956, virtually all private industrial establishments had been transformed into joint operations. At the same time, 400,000 private commercial establishments became joint enterprises, and another 1.4 million units became marketing cooperatives.

The process of "socialist transformation" of industry and commerce was accompanied by a policy of paying interest to the private owners of joint enterprises at a fixed rate. Private shares of capital were assessed,[6] and interest payments based on the assessed value were made to the private owners, generally at a rate of 5 per cent per annum regardless of the profits or losses of their enterprises. With such an arrangement of "buying off" private shares, the capitalists were forced to give up whatever control they had retained over their enterprises. Interest payments to former owners were originally scheduled to expire in December, 1962, but were later extended to 1965.[7] With the cessation of interest payments, the capitalist class in China was eliminated both in theory and in fact. All of the joint enterprises became state enterprises.

The incorporation of traditional small production units, including farmers, small traders and peddlars, and individual craftsmen, into cooperatives represented another important Communist goal in the overhaul of the economic structure. The farm sector, which was and still is the most important sector of the Chinese economy, underwent a process of transformation much more thorough and complicated than other sectors.

In the early 1950's, there were in China some eight million independent small craftsmen.[8] Communist policy toward these craftsmen was first to organize them into small "supply and marketing groups," which purchased raw materials from and sold finished products to state commercial establishments. Group members continued to own their equipment and to produce as independent units, accounting for their own profits and losses. The supply and marketing groups were gradually expanded into a more advanced form of handicraft cooperation, known as "supply and marketing cooperatives," in which members not only joined together to acquire raw materials and sell products to state commercial establishments, but also cooperated in the manufacturing process. A portion of the profits arising out of their joint operation were used to purchase communally owned implements. At the same time, the members began to turn their own implements over to the cooperative to form a common pool of capital. Finally, handicraft cooperation reached its most advanced form, the handicraft producer cooperative,

---

6. At the end of 1956, total private shares in the state-private jointly operated enterprises were valued at only 2.5 billion *yuan*.

7. Total interest payments were about 120 million *yuan* each year, and the number of recipients, 1,140,000.

8. In addition, there were 12 million peasants engaged in commercial handicraft production on a part-time basis, while a large portion of the peasant population produced handicraft items for their own consumption. There were also a large number of persons working in handicraft workshops.

in which all means of production became the property of the cooperative. The management of this type of enterprise was unified, and profits and losses were shared by the members. Income, after deductions for taxes, reserves, and welfare funds, was distributed to members in wages and bonuses according to government regulations. By the end of 1956, nearly 92 per cent of all individual craftsmen in China had been absorbed into some form of handicraft cooperation.

Small traders and peddlers also were gradually drawn into cooperatives. In the mid-1950's, there were about 2.8 million such small trading units, employing 3.3 million people and accounting for 65 per cent of the trade volume handled by private merchants. Small traders and peddlers were at first organized into cooperative groups which acted as retail distributors, commission agents, or purchasing agents for state commerce. Members acquired merchandise from the state and sold it at official prices. Their income was derived from the difference between wholesale and retail prices, or from a commission paid by the state. A more advanced form of trade cooperation was the cooperative store, in which profits and losses were shared among the members. By the end of 1956, 1,150,000 small units in retail and catering trades, 46 per cent of the total number, had been organized into cooperative groups, while 800,000 units, 32 per cent of the total, were merged into cooperative stores. The remaining units were too widely scattered throughout the country to be easily made cooperatives. These units remained on an individual basis under the close supervision of state commercial organizations.

In the traditional Chinese agricultural system, the majority of the farming population remained at the margin of subsistence while the bulk of agricultural savings was in the hands of landlords and rich peasants. The latter were considered by the Communists to be politically unreliable. Immediately after the seizure of power, therefore, they launched a vigorous land-reform campaign to eliminate the wealthier landed groups.

The tactics employed involved "relying on the poor peasants, uniting the middle peasants, neutralizing the rich peasants and eliminating the landlord class" (C. Y. Cheng, 1963, p. 24). It was a firm belief of the Communist leadership that in a country with a large farm population, the peasant masses had to be won over until the power of the regime was fully consolidated. To gain good will, the land expropriated from the landlords was redistributed to the poor peasants, including tenant cultivators and landless laborers.

The middle peasants, who were relatively well endowed with managerial and farming skills, with implements and draught animals, were to join the poor peasants in an effort to restore agricultural production to pre-war levels. The rich peasants were allowed to retain the land directly cultivated by them and their hired labor. This was a policy significantly different from the earlier Communist agrarian program, which called for the elimination of the rich peasants together with the landlords. The change in the policy toward the rich peasants, as Liu Shao-chi explained, was necessary not only to maintain order in rural areas and expedite agricultural recovery by taking advantage of the higher productivity of the rich peasants, but also to isolate the landlords (*Agrarian Reform Law*, 1959, pp. 59–85). The main goal of the initial program of land reform was not to abolish private ownership but to eliminate the landlord classes.[9]

The land reform was largely completed by the end of 1952, when 700 million *mou* of land were redistributed among 300 million poor and landless peasants (Hsu, 1959, pp. 117–18). The rural economy that emerged was one of peasant proprietors with average holdings even smaller and more scattered than before.[10] The small peasant landowners cultivating tiny holdings by indigenous methods and living at the margin were not capable of supplying the surplus food and raw materials necessary for industrialization. The peasants were tempted to raise their living standards and were unwilling to respond to pleas for self-sacrifice. In the eyes of the Communists, once in possession of the land, the peasants became petit-bourgeois and ceased to be a revolutionary force. From the standpoint both of state control and of production efficiency, the small peasant economy could not be tolerated for long. As soon as they felt their political position reasonably secure after the abolition of landlordism, the Communists began a series of measures aimed at agricultural cooperation and the eventual elimination of all private ownership in the countryside.

The first step toward cooperative farming was the organization of peasant households into *ad hoc* mutual aid teams, which were gradually converted into permanent units. Private ownership of land was retained by each household, which continued to be an independent production unit. Labor, draught animals, and implements were pooled on a cooperative basis. The *ad hoc* mutual aid team, usually comprising three to five households, was organized at peak seasons to perform the major tasks. After the work was completed, the team was dissolved and accounts among member households were settled.

The permanent team was a more complicated arrangement maintained continuously throughout the year, usually consisting of six to ten households. Members cooperated not only in farm work during the busy season but also in subsidiary activities in the slack months of winter. A small communal sinking fund was established to build up a collectively owned stock of implements and draught animals. The Communists viewed the permanent mutual-aid arrangement as "embryonic socialism." Some 45 million peasant households, representing 40 per cent of the total number, were organized into 8.3

---

9. The prefatory article of the *Agrarian Reform Law* of June, 1950, states explicitly: "The land ownership system of feudal exploitation by the landlord class shall be abolished and the system of peasant land ownership shall be introduced in order to set free the rural productive forces, develop agricultural production and thus pave the way for New China's industrialization." *Ibid.*, p. 1.

10. Upon completion of land reform the average size of landholdings for the country as a whole was about two acres. It was larger than the average in the North, Northeast and Northwest, and smaller in the South and Southwest. See Hsueh, 1960, pp. 98–101.

million mutual aid teams by the end of 1952 (State Statistical Bureau, 1960, p. 34). Two years later, the figure had increased to 60 per cent (Hsu, 1959, p. 130).

As soon as the majority of the peasant households had joined mutual aid teams, the Communists began to urge them to form loose cooperatives, known as "agricultural producer cooperatives of the elementary type."[11] These cooperatives differed from the mutual aid teams in that land was pooled along with labor, draught animals, and implements. While the member households retained title to the factors of production, their use was controlled and planned collectively by the cooperative management committee on the basis of targets received from the government. To accumulate a capital stock, each cooperative allocated a certain proportion of current net income to a reserve fund. After the allotment to the reserve fund and the payment of taxes, the remainder of the income was divided into a land share and a labor share for distribution among the member households. The land share was based on the quantity and quality of the land holdings, the draught animals, and the implements which each member contributed to the common pool, while the labor share was divided according to workdays or work-points.[12]

A small private plot was allocated without charge to each household, according to the size of the household and the amount of arable land in the village. The peasants were allowed to grow vegetables and other garden crops on the private plots on their own time and to dispose of these crops by themselves. But the overwhelming proportion of the resources was controlled by the cooperatives.

By the summer of 1955, the number of peasant households in cooperatives of the elementary type had reached 16.9 million, about 15 per cent of the total. In October of that year, the Communist Party decided to accelerate the tempo of agricultural cooperativization. By the end of 1956, more than 110 million peasant households, 92 per cent of the total, had been grouped into cooperatives.

Simultaneously, agricultural collectives, officially termed "agricultural producer cooperatives of the advanced type," began to appear. Collectives were larger than the cooperatives and were marked by common ownership of practically all major peasant properties, including land. Only domestic livestock, small implements, and scattered trees were left in private hands. Payments to the previous owners for the use of their land, draught animals,

11. These cooperatives had appeared in China before 1955, though on a very small scale. It was reported that the number of peasant households joining the elementary cooperatives was 187 in 1950, 1,588 in 1951, 57,000 in 1952, 273,000 in 1953 and 2,285,000 in 1954, while the total number of peasant households was about 125,000,000. State Statistical Bureau, 1960, pp. 34–35.

12. Under the work-day system, each job was given a "norm" determined on the basis of the performance of an "average" worker in a working day. Fulfillment of the norm for each job entitled the peasant to a number of work-points assigned according to the skill and strength required and the importance of the job. Usually, one workday was equivalent to ten work-points. "Model-Regulations for Agricultural Producer Cooperatives," 1955, pp. 141–149.

and implements were discontinued. The workday system remained the essential method of distributing the product, but a contract system, under which a task was assigned to a production team in the collective at a fixed price, was sometimes used. Only 4 per cent of the peasant households had joined collectives at the end of 1955, but the figure rose to 88 per cent a year later (State Statistical Bureau, 1956, pp. 63–65).

By June, 1957, 93 per cent of the peasant households had been collectivized (Hsu, 1959, p. 133). The remaining households, largely in border regions or national minority areas, remained in some form of cooperative. Agricultural collectivization in China was virtually completed ten years ahead of the original schedule.[13]

There were both similarities and differences between the Chinese and the Soviet collectivization programs. The similarities were particularly evident in the structure of the collective farms in the two countries. The Chinese collective was patterned closely on the Soviet *kolkhoz* model. The system of management, the method of distributing the product, the provision of private land plots, and private ownership of a limited amount of other property in the Chinese collectives did not differ substantially from their Soviet counterparts. But there were also significant differences, the most striking being the relatively non-violent and rapid transformation from private ownership to complete collectivization in China. The Chinese Communists studied the Soviet experience closely and were able to avoid the disturbances which attended Soviet collectivization.[14]

During the prolonged period of civil war in China, the Communists had ample opportunity to experiment with different types of land reform tactics in the areas under their control, and to train cadres in implementing these tactics. After 1949, they were able to send the cadres with successfully tested techniques to organize the peasants in a nationwide campaign against the landlords. In the Soviet Union the initial revolutionary process of land reform had been spontaneous, carried out by the peasants themselves with scarcely any Communist control. With the removal of the landlords, varying forms of farm cooperation were immediately instituted in China, to take land away from the peasants before they could consolidate their newly won gains.

Compared with the Soviet Union, the process of collectivization in China was not only smooth but rapid. It took the Chinese only five years (from 1953 to 1957) to move from private landholding to complete collectivization. There was a bitter debate within the Party on the speed of collectivization, and the group led by Mao favoring a big push evidently won. The result was an all-out effort, particularly in 1956, during which 84 per cent of the peasant households were brought into collectives.

13. The first document which outlined the precise schedule for the socialist transformation of Chinese agriculture appeared to be *The First Five Year Plan* published in 1955. According to the *Plan,* collectivization in China was to be completed in fifteen years, that is, by 1967.

14. Soviet collectivization encountered strong resistance from the peasants and had many deleterious consequences. In 1930–34, agricultural production declined sharply, half of the livestock were slaughtered, and many peasants died of hunger.

The decision to speed up collectivization was probably made because, during the first two years of the First FYP, agriculture failed to yield a surplus sufficient for the support of an ambitious industrialization program. In 1953 and 1954, grain production increased only by 1.5 per cent and 2.3 per cent respectively, while cotton production actually fell by 9.9 per cent and 9.4 per cent in the same years. In addition, the targets for silk, tea, jute, oil crops, and livestock were not fulfilled (Liao, 1955, pp. 10–23). The underfulfillment of the agricultural plan targets and the ability of the peasants to evade delivery of quotas jeopardized the newly established system of compulsory deliveries instituted to extract an agricultural surplus. The problem was aggravated by the exhaustion of Soviet credits and the need to repay the accumulated debt, which meant stepped up deliveries of agricultural produce to the Soviet Union. The solution to these difficulties was collectivization, through which, it was hoped, both output and savings could be expanded.

Agricultural collectives did not last long. After one year of existence they were superseded by new and much larger organizations known as people's communes. The commune movement got underway in June, 1958,[15] and immediately swept the country. By September of that year, the total number of communes was said to have reached 26,425, containing over 120 million households, or 99 per cent of the peasant population (*Statistical Work*, No. 20, 1958, p. 23). The average number of households in a commune was about 4,600, in contrast to an average of 158 households in a collective (C. Y. Cheng, 1963, p. 39).

The communes were not merely much larger than the previous collectives; they also entailed a far higher degree of collectivization. The separate structures of government administration and agricultural collectives were brought together. The new unit took over all the administrative functions of the *hsiang*, previously the lowest unit of local government. Unlike the collective, which was primarily agricultural in function, the commune also engaged in industry, trade, banking, education, and cultural work. Moreover, the commune had its own militia and political organizations.

The entire way of life was much more collectivized in the commune than in the previous collective. Mess halls, nurseries, kindergartens, and homes for the aged were established. The traditional family system was marked for eventual elimination. Private ownership was all but abolished. Private plots, houses, trees, orchards, fishponds, and livestock, which individuals had previously been allowed to retain, were now collectivized.

The commune normally consisted of a number of brigades, which were themselves subdivided into production teams. The brigade was the basic unit in charge of production activities under the direction of the commune management committee, and had its own profit and loss account which was then merged into that of the commune as a whole. The production team was the basic unit of labor organization.

15. The first commune, known as the "sputnik commune," was organized in April, 1958 in the Sinyang district of Honan Province.

Distribution was governed by a "part wage, part free supply system." Members received a free supply of food in the mess halls and a monthly payment of wages determined on the basis of work attitudes, physical strength, and technical skills. The new system was in reality not very different from the distribution methods employed in the cooperatives and collectives. The most significant change under the commune was shifting of the base for calculating wages from the income of the cooperative or collective to that of the entire commune.

What motivated the Chinese to establish the commune system is not clear. The reasons officially given were two: "The all-round, continuous leap forward in China's agricultural production; and the ever-rising political consciousness of the 500 million peasants" (*Communist China*, 1955–59, 1962, p. 454). Neither reason is convincing. The harvest of 1958 was very good, primarily because of exceptionally favorable weather. There was no evidence to support the view that the "political consciousness" of the Chinese peasant was rising. On the contrary, there were many reports of the deficiencies of agricultural collectives and of peasant discontent.

From an economic point of view, the commune system seemed to be a formula for solving the conflict between the desire for rapid industrialization and the low level of per capita agricultural output. This had bothered the policy makers for many years, but their attempt in 1955–56 to resolve it through agricultural collectivization was not successful. Over-all agricultural production did not gain significantly in 1956, and in 1957, the output of all crops reported, except for grain, cotton and sugar cane, declined (N. R. Chen, 1967, pp. 338–39).

The difficulties encountered in agricultural tax collection and crop procurement, which motivated Mao and his followers to favor a big push in 1955, were not alleviated. On the contrary, the problem became critical in 1957 when many collectives concealed their output by maintaining two sets of books, one recording actual output for internal use, and the other listing fictitious figures for submission to the government (C. Y. Cheng, 1960, p. 338). On the eve of the Second FYP, scheduled to begin in 1958, the planners were confronted with a dilemma: a high rate of industrial investment could not be maintained if agriculture continued to stagnate, but agricultural output could not be raised without a significant curtailment of investment in industry.

The Chinese apparently reached the conclusion that collectivization along Soviet lines could not be relied upon to resolve this dilemma. A new approach, the Great Leap Forward, was adopted. As we have already indicated, the Great Leap continued to emphasize heavy industry, without diverting investment resources from industry to agriculture. For the countryside this involved mass mobilization of presumably underemployed rural labor primarily in three ways: (1) large scale construction projects in which human labor constituted the bulk of the input, such as irrigation, flood control, and land reclamation; (2) greater application of labor to increase unit land yields by close planting, deep plowing, and the like; and (3) small plants using labor-intensive techniques. The commune was instituted as a means of implementing this strategy; it was believed that a larger unit of organization would face fewer difficulties than the smaller collective in allocating labor and managerial skills to staff the new projects.

Instead of moving the economy forward, the policies of the Great Leap Forward caused a deep economic crisis. Agricultural production suffered a marked decline in 1959 despite initial claims to the contrary. When the harvests failed repeatedly during the next two years, severe food crises developed. The sharp downturn in agricultural output led to a cumulative decline in industrial production and in the national product; agriculture became the critical bottleneck to the further development of industry.

The causes of the farm crisis were numerous. The official explanation cited three consecutive years of natural disasters, shortcomings of the cadres, and sabotage. But these reasons provide only part of the explanation. There are others, two of them fundamental. First, technological constraints on the expansion of output were overlooked. The labor utilization scheme was pushed beyond a rational limit, resulting in negative marginal productivity of labor. Outstanding examples were "that deep plowing had broken the water table of paddy fields necessitating the use of much more water than before, that the reckless digging of irrigation canals had alkalized the soil and close planting exhausted it, and that ignorant cadres intent only on enforcing official directives, had ordered sowing it the wrong time or place" (*Far Eastern Economic Review, 1962 Yearbook*, p. 58).

Second, and perhaps even more important, the planners neglected the importance of material incentives, and attempted to expand production by organizational and ideological means alone. Under the commune system, individual interests were de-emphasized. Private plots were confiscated and rural free markets abolished. Differentiated rewards were replaced by egalitarian wage systems. In the initial phase of the communes, wages were calculated on the basis of the income of the entire commune rather than that of the collective or the production team. In consequence, the relationship between the reward received by the individual peasant and the amount and quality of his labor became more remote and obscure than in the collective or the production team. The lack of personal incentives soon had a negative effect on labor productivity. It was admitted that a portion of the autumn harvest in 1958 was damaged by peasant sabotage (C. Y. Cheng, 1963, p. 45).

As peasant discontent grew, the regime was forced to take corrective measures. The first indication of a retreat was the Party resolution of December, 1958, in which the commune system as a transitional stage to full communism was de-emphasized and the importance of personal incentive stressed (*Communist China, 1955–59*, 1962, pp. 490–503). But concrete measures were not taken until the summer of 1959, when grain rations were distributed to individual households instead of to mess halls, and the percentage of food distributed as free supply was reduced (C. Y. Cheng, 1963, pp. 46–47, and Walker, 1965, p. 17). Free markets in rural areas in the form of trade fairs were gradually re-introduced (Perkins, 1966, pp. 91–95). By the summer of 1960, small private plots had been returned to individual peasants.

Finally, measures were adopted to transfer control of production and income distribution down to lower levels in the commune organization. Ownership, profit and loss calculation, and wages were made the responsibility of production brigades early in 1961, and of the production team a year later. By 1963 the part wage, part free supply system had been completely replaced by the methods of distribution previously used in the cooperatives and collectives. After four years of experimentation with the commune, the basic unit of production and income distribution in Chinese agriculture returned to its pre-1958 structure.

Mobilization of Savings

One of the major purposes of the Communist regime in reorganizing the economy was to facilitate the mobilization of savings for fueling the development process. Table VI-1 presents estimates of the rate of investment for

TABLE VI-1
PRE-WAR AND POST-WAR RATES OF INVESTMENT IN CHINA

Period	Gross capital formation (billion 1952 yuan)	Gross domestic product (billion 1952 yuan)	Ratio of investment (per cent)
1931–36	4.77	63.50	7.5
1952–57	20.82	86.94	24.0
1958	26.94[a]	117.71	22.9[a]
1959	32.00[a]	127.98	25.0[a]

Note:
[a] State fixed investment.
Source:
Yeh, 1968.

prewar and post-war periods. During the years 1931–36, a relatively normal period, the gross investment ratio was 7.5 per cent. If schemes of forced savings had not been imposed, the rates of savings in the Communist period would probably not have been much higher than those of the prewar years, since per capita income had not increased substantially. During the First FYP, however, the investment rate rose to 24 per cent. In the years of the Great Leap Forward, the ratio of state fixed investment to gross domestic product was 23–25 per cent. (These figures may be overstated due to the already mentioned overvaluation of producer goods).

Since available external sources for financing Chinese economic development were negligible (Mah, 1961, pp. 33–48), the high investment rates in the post-war period were made possible primarily through the mobilization of domestic savings and their conversion into capital investment. The bulk of investment was financed from sources directly under state control. A computation of the percentage distribution between state and non-state gross investment during 1952–57 indicates that the state-financed portion increased from 70 per cent of the total in 1952 to 88 per cent in 1957 (Yeh, 1968, Table B–1).

State investment in China is financed through a budget which is national in scope, covering both the central and local governments. Since budgetary revenues from all sources are used to finance total budgetary expenditures, including investment, it is impossible to show from which specific revenues state investment came. However, since investment accounted for a large proportion of total budgetary expenditures,[16] an examination of the major sources of budgetary revenues helps explain how state investment is financed.

Table VI–2 presents the sources of Chinese budgetary revenue for selected

TABLE VI–2
SOURCES OF STATE BUDGETARY REVENUE IN CHINA, 1950–59
(millions of current *yuan*)

	1950	1952	1957	1959
1. Tax revenue	4,898	9,769	15,490	20,470
a. Industrial and commercial taxes	2,363	6,147	11,300	15,698
b. Agricultural taxes	1,910	2,704	2,970	3,300
c. Salt taxes	268	405	620	650
d. Customs duties	356	481	460	650
e. Miscellaneous tax receipts	1	32	140	172
2. Non-tax revenue	1,621	7,791	15,530	33,690
a. Profits from state enterprises		4,653	11,363	28,590
b. Depreciation reserve and other income from state enterprises	870	1,077	3,057	4,770
c. Foreign loans	244	1,305	23	0
d. Domestic loans	260	0	650	0
e. Other receipts	247	756	437	330
3. Total budgetary revenue	6,519	17,560	31,020	54,160

Source:
Ecklund, 1966, p. 20.

years during the period 1950 to 1959. Revenue rose by more than eight times during the period. Budgetary revenue increased from 10 per cent of gross national product in 1950 to over 30 per cent in 1958, and even higher in the following year.

Taxes accounted for more than half the total revenue until 1957, when non-tax revenue began to exceed tax revenue. The most important items

16. During the First FYP, for example, state investment constituted 30.3 per cent of total budgetary expenditures in 1953, 30.4 per cent in 1954, 32.3 per cent in 1955, 44.3 per cent in 1956, and 42.2 per cent in 1957.

of tax revenue were industrial and commercial taxes, which included commodity taxes and taxes on the net income and gross receipts of enterprises. They constituted 48.2 per cent of total tax revenue in 1950 and 76.7 per cent in 1959. The agricultural tax, which was imposed on gross farm output on the basis of "normal" rather than actual yield, was the second most important source of tax revenue. In 1950 it provided 39 per cent of tax revenue, but its share declined to 16 per cent in 1959. Salt taxes and custom duties, both of which were important in traditional China, continued into the Communist period. Like the agricultural tax, these two revenue sources also declined in importance relative to total budgetary revenue.

The decline in the relative importance of the agricultural tax by no means implied a reduction in the tax burden levied on the peasants. The peasants had to pay a higher tax in the form of "profits from state enterprises", which made up roughly 42 per cent of total budgetary receipts during the First FYP, and more than half of total revenue in 1959. The state procurement scheme for agricultural commodities, which was adopted for grain in 1953 and for cotton and other major crops in subsequent years, required the peasants to sell the government a specified amount of their products (after the payment of agricultural taxes) at relatively low prices. They then had to purchase manufactured commodities and even their own grain back at higher prices. The difference between the purchase and sales prices was channeled into the state budget partly as commodity taxes and partly as profits from state trading enterprises.[17]

Apart from profits from state enterprises, the category "depreciation reserves and other income from state enterprises"[18] was the most important of the non-tax revenues. The bulk of the income in this category came from depreciation reserves. As a proportion of total budgetary receipts, depreciation reserves rose from 5.6 per cent in 1952 to 7.3 per cent in 1959, and averaged more than 6 per cent over the entire period (Ecklund, 1966, p. 20). Since a great deal of industrial equipment was newly installed during 1952–59, depreciation reserves far exceeded the outward flow of expenditures for replacing fixed assets actually scrapped. Therefore, a large proportion of the depreciation reserves served as a source of funds for new investment rather than as a commitment from the state budget for replacement purposes.

Other categories of non-tax receipts were domestic and foreign loans. Domestic loans took the form of government bonds. The first national bond issue was the People's Victory Bond in 1950. From 1954 through 1957,

---

17. According to one Chinese article, "an industrial product is subject to taxation many times from the point of production to the point of final sale, and some intermediate goods in the process of continuous production are also taxed." Wang and P'eng quoted in Li, 1959, p. 150.

18. Included in "other income from state enterprises" are "return of surplus working capital," "receipts from the sale of fixed assets," and "income from other business activities" including proceeds from experimental products of research agencies, the income of broadcasting and news services, and the income of surveying and planning activities.

National Economic Construction Bonds were issued each year. National bonds were discontinued in 1959, when provincial governments were permitted to issue local economic construction bonds. The amount of revenue derived from government bonds was not very significant. During 1950-58, domestic bonds brought in 3,762 million *yuan*, about 1.8 per cent of the total budgetary revenue during that period (Ecklund, 1966, p. 20).

Foreign loans came solely from the Soviet Union. During 1950-57, total Soviet loans to China amounted to 5,294 million *yuan*.[19] A third of this total was used for the purchase of capital goods from the Soviet Union, with the remainder not clearly identified but probably used mainly for military equipment. Receipts of Soviet credit accounted for 3.7 per cent of total budgetary revenue in 1950, 7.4 per cent in 1952, and only 0.1 per cent in 1957.

A Western estimate indicates that during the First FYP the peasants had to bear the heaviest burden of net domestic revenue (48 per cent), followed by urban residents (37 per cent), and private businessmen (12 per cent) (Mah, 1961, p. 45). Communist sources appear to confirm this finding; one of them states that "during the first plan period over 50 per cent of the revenue had come either directly from the agricultural sector, or from industrial production, commerce, foreign trade and transportation, which are closely related to agriculture" (Kiang Tung, quoted in Mah, 1961, p. 467).

As already noted, capital investment was financed in the main from state resources. In gross terms the share of non-state investment in total investment declined from 30 per cent in 1952 to 12 per cent in 1957. The relative importance of non-state investment may have decreased even further in subsequent years due to the programs of the Great Leap and the advent of the commune program. In absolute terms, however, the amount of non-state gross investment was quite stable, remaining between two and three billion 1952 *yuan* during 1952-57 (Yeh, 1967, Table B-1). Roughly two thirds of non-state investment was devoted to agriculture, the remainder to industry, traditional transportation and housing.

An interesting question arises: was non-state investment financed through voluntary savings? In the absence of quantitative information the question cannot be answered with any degree of certitude. Available evidence seems to indicate, however, that at least part of non-state investment was the result of forced savings measures. In farm and handicraft cooperatives and in private enterprises, for example, a specified proportion of income was required to be retained for capital accumulation. Frequently, the proportion was set so high that part of the income alloted to investment would otherwise have been used for consumption purposes. In addition, state and cooperative credit increased substantially during the1950's (Perkins, 1965, p. 79). Peasant savings also rose to some extent as a result of the introduction of mobile

---

19. Ecklund, 1966, p. 92. In 1961, following the Great Leap failures, the Soviet Union provided China with 320 million dollars of outstanding short-term indebtedness over a 5-year period, and also extended to China a loan of 46 million dollars for the importation of 500,000 tons of sugar from Cuba.

banks in the rural areas, new types of savings deposits, new financial institutions, and savings through insurance contracts.

National Economic Planning

Reorganization of the economic structure and mobilization of savings were the major steps taken by the Communists toward state control over the economy. Resources diverted to the production of food, clothing, and other consumer goods were kept at a level barely sufficient to sustain the productive energies and morale of the population. The bulk of investment was devoted to producer goods, military hardware, and the development of nuclear capability.

To ensure the desired product mix required the institution of national economic planning. As soon as the Communists took over the economy, a central planning apparatus was set up to map out a scheme for economic recovery. But long term, unified planning did not exist until 1953, when the First FYP began. The Second Plan, which was scheduled for 1958–62, had to be abandoned because of the Great Leap. Economic plans for these and subsequent years were apparently prepared on an annual *ad hoc* basis, although no details were published. In 1965, Peking announced that a Third FYP was being prepared for 1966–70. Like the Second Plan, however, the Third Plan was not put into effect; this time, the cause was the chaos created by the Cultural Revolution. There has been therefore only one complete five-year plan period, the First FYP, as far as is known outside China.

The national plan was composed of a set of component plans for industrial production, agricultural production, transportation, labor and employment, material allocation, commodity flow, capital construction, foreign trade, production costs, commodity prices, technological development, and social, cultural and welfare undertakings (Wu, 1967, pp. 99–119). State enterprises were the units responsible for carrying out the directives contained in the plan. In industry, the directives were transmitted to the enterprise through administrative channels in the form of control figures or targets. Included in the targets were gross value of output, volume of output, product mix, product quality, costs, profits, employment, wages, and labor productivity (Kwang, 1966, pp. 61–99). The enterprise formulated its plan on the basis of these targets, and was required to fulfill its output and profit plans and to lower its cost of production.

In agriculture, the procedure was not as elaborate. Output targets for major crops were calculated in terms of requirements and of availability of inputs, and were then assigned by the central government to the provinces, by the provinces to the counties, and by the counties to the cooperatives or communes. There was none of the detailed supervision found in the case of industrial production. The assigned output targets were accepted by the cooperative or commune as reference goals, but not necessarily as firm quotas which it had to fulfill.

The quality of agricultural planning was therefore inferior to that of industrial planning. With the exception of state farms, which accounted for a very minor portion of total agricultural output, agriculture remained outside the government sector, so that the state lacked direct control over the allocation of resources within agriculture. There were about 120 million peasant families working in a very large number of production units, with different farming techniques and under widely varying soil and weather conditions. Detailed supervision of agricultural production was obviously out of the question. Moreover, the uncertain nature of agriculture itself made the calculation of firm targets impossible. Every year there is flood or drought in some part of the country. A deviation of plus and minus 10 per cent of agricultural output targets was allowed for in the annual plans during the First FYP (Government of India, 1956, p. 78).

Information is the lifeblood of economic planning. In order to prepare a vast planning program, determine its practical aspects, control its implementation, and evaluate its performance, the Chinese government must maintain a statistical system containing detailed records of past performance and a continuous flow of information on work in progress in every part of the economy. As early as 1950 the regime conducted a number of special surveys; the results provided the basis for the state plans of economic rehabilitation and reconstruction in the first three years. With the establishment of the State Statistical Bureau in October, 1952, a unified system of regular statistical reporting was introduced. All production units in the country were required to fill in statistical schedules issued by the Bureau and to submit them regularly. Statistical terms and methods of computation were standardized.

The statistical fiasco of the Great Leap Forward period caused the quality of Chinese data to deteriorate considerably. When the Party cadres and the "masses" took over part of the statistical work, the very existence of the state statistical system was threatened. Chinese statistics published in 1958 and 1959 were in a great many cases exaggerated or distorted. Since 1960 the Chinese have ceased to publish either statistical information or material on the statistical system itself, so we have no way of knowing how reliable economic data are at present.

The government set up a State Planning Committee in 1952, and two years later reorganized it into the State Planning Commission. It was charged with responsibility for the preparation of both long-term and annual economic plans. At the same time, a State Construction Commission was established to assume charge of capital construction work.[20] In May, 1956, a new State Economic Commission was created, taking over from the State Planning

---

20. In February, 1958, the State Construction Commission was abolished and its functions were taken over by Planning and Economic Commissions and by the Ministry of Building. A State Capital Construction Commission was established in October, 1958, and abolished in 1961 after the Great Leap. The Commission was restored in 1965 when the government decided to start the Third FYP.

Commission the responsibility of formulating the annual plans within the framework of the five-year plan prepared by the Planning Commission. The Economic Commission was also vested with authority to coordinate the work of individual government agencies and enterprises in order to maintain a balance in the allocation of materials.

The functions of the State Planning Commission were thus eventually confined largely to preparing perspective, long-term and five-year plans. Planning organizations also existed at the provincial and county levels under the overall direction and guidance of the State Planning Commission. These sub-commissions were set up in all of the 23 provinces, four autonomous regions (excluding Tibet), and three independent municipalities (Peking, Shanghai and Tientsin). Each of the 2,117 counties throughout the country also had its own planning department.

The first major step toward plan formulation was the issuance by the State Planning Commission of control figures: targets or quotas expressed in both physical and value terms, input and cost coefficients, and other general and aggregative indicators governing various aspects of economic activity. The control figures were worked out by the Commission on the basis of the political directives of the Communist Party and the quantitative data supplied by the State Statistical Bureau, in consultation with the ministries concerned.

After approval by the State Council, the control figures were sent down to lower levels of the government through two channels. One passed through the central government ministries and their counterparts in the provinces and counties. The other channel went through the planning commissions at the provincial and county levels. On the basis of these control figures the various lower echelons drew up their own draft plans, which were then submitted to the State Council and the State Planning Commission. A draft plan for the economy as a whole was then compiled by the State Planning Commission. After approval by the State Council, it was sent to the National People's Congress for approval. The final plan was then transmitted downward to the lower echelons through the two channels mentioned above.[21]

The preparation of the annual plans during the First FYP period, especially

---

21. The steps involved in formulating annual plans may be seen from the time schedule for preparing the 1956 plan:
July–August, 1955. Preparation of control figures.
August, 1955. Issue of control figures by Planning Commission.
August–October, 1955. Preparation of draft plans by ministries, provinces, municipalities, enterprises.
End of October, 1955. Ministries and provincial governments send their plans.
November–December, 1955. Planning Commission formulates the draft plan.
Mid–end January, 1956. Planning Commission makes necessary revisions in the original draft plan in the light of discussions at the conference.
Mid-February, 1956. Draft plan printed and submitted to State Council.
End-February. State Council approves the annual plan.
Industrial enterprises generally did not receive their plans until the end of March. Until they received the final plan, they were allowed to work on the basis of a provisional plan.

for the years from 1955 through 1957, was generally based on the procedure outlined above. With the advent of the Great Leap in 1958-59, national economic planning became chaotic. The frequent revisions of major industrial and agricultural targets during these years made planning almost impossible. The post-Leap economy went through stages of stagnation, readjustment, and recovery with, as far as we know, no long-term economic planning. The Cultural Revolution undoubtedly dealt another severe blow to the operation of the planning system. Chinese sources indicate that the Third FYP was attacked by Mao Tse-tung's followers as "revisionist" and "capitalist" in its approach. An economist who was reported to be instrumental in the drafting of the Third FYP was denounced as the "Liberman of China," and was evidently purged.[22]

The key unit in implementing the national economic plan was the enterprise, an independent economic accounting entity. The plan targets for each enterprise were set by the state. The basic responsibility of the enterprise was to translate these targets into orders and directives for its subordinate units. The state judged plan performance on an enterprise basis.

The implementation of the enterprise plan was controlled by physical and financial targets. On the physical side, output mix and input requirements for each enterprise were specified by the central planning authorities. The key element in physical control was the centralized allocation of major materials and machinery. Output and input targets were set in such a way that all resources allocated to the enterprises had to be utilized efficiently if the targets were to be fulfilled.

On the financial side, a number of targets, including profits, cost reduction, and the wage bill, were designed to back up the physical targets. The key feature of the financial measures was centralized control over sources and uses of enterprise funds. This control took two forms. One comprised the sales and profit taxes which took away a large portion of the gross income of the enterprise. The other consisted of elaborate rules governing the use of working capital and loans. The working capital of a state enterprise was financed out of the state budget, while loans could be granted to the enterprise by the state when working capital was insufficient to meet needs. The People's Bank of China handled the payment of working capital and granted loans. It was also authorized to supervise expenditures of the enterprise to ensure that they were made in compliance with the central plan.

In theory, the enterprise was required to prepare detailed income and outlay plans, upon which the grant of working capital and loans was based. The freedom of the enterprise to use these funds was severely limited, however. For example, the use of working capital for fixed investment purposes or for raising the wage bill beyond the target was prohibited. In practice there were

---

22. The economist is Sun Yeh-fang, who in the 1950's served as Deputy Director of the State Statistical Bureau and Vice Chairman of the State Planning Commission, and had been director of the Economic Research Institute of the Chinese National Academy of Sciences until the Cultural Revolution.

many instances in which the financial regulations were not strictly observed by enterprises (Perkins, 1968, p. 620).

The Use of Markets and Prices in Resource Allocation

The effectiveness of the centralized system was greatly hampered by the level of China's economic development. The large number of industrial firms, most of them small in size and backward in the techniques of production, made centralized control anything but a simple task.[23] It was therefore necessary to adopt decentralized measures, including reliance on markets and prices, to supplement the system of centralized control.

An outstanding example was the maintenance of high prices for producer goods in the early 1950's. During 1950–52, wholesale prices of industrial commodities were to a large extent determined by market forces. Major industrial products, including cement, coal, and steel products, were partly sold on the market. This practice was permitted because in those early years the regime was unable to achieve the necessary control over allocation. The prices of producer goods in these years rose relative to other prices (such as consumer goods prices, agricultural prices, wage rates, and transportation charges), primarily due to an increase in demand resulting from industrial rehabilitation and the Korean War. With the introduction of the First FYP, the state began to take steps to control markets for producer goods. The prices for these goods were frozen at the levels prevailing during the Korean War, which were substantially above average cost of production (Perkins, 1966, p. 110).

One of the purposes of the planners in maintaining relatively high prices for producer goods was to help ration them, instead of relying entirely on the network of centralized material allocation. Thousands of industrial commodities were produced, and only a very small proportion of them were directly rationed by the central planning authorities.[24] The industrialization drive had caused the demand for most industrial commodities to exceed supply, and high prices were considered necessary to prevent wasteful use.

Another example of the attempts made by the planners to use markets in allocating resources may be found in the exclusion of some portion of industrial output from the province of centrally planned commodity balances. Firms used a wide variety of production methods for the same product. In consequence, there were large disparities in average cost or in inputs per unit of output among firms in the same industry. It was almost impossible for the central planners to coordinate efficiently the resources employed by these many firms. Efforts were made to rely principally on markets and prices

23. A large portion of industrial output was produced by individual craftsmen and by factories employing fewer than 15 persons. In 1953 there were 176,000 firms, of which 145,000 were small-scale enterprises employing fewer than 15 persons. The small-scale enterprises produced nearly one-third of the gross value of industrial output.

24. In 1956, for example, there were only 235 industrial commodities rationed by the central planners, and not even these could all be distributed in a systematic way.

to coordinate the work of small firms, particularly those producing consumer goods, although the central planners still provided general guidance.

Other decentralized measures included transfer of administrative duties from higher to lower-level decision makers. In 1958, the responsibility for directing a large number of enterprises was transferred from the central government to local governments, except for those enterprises producing important materials. At the same time, local governments were vested with authority to determine prices for many commodities, particularly those produced locally.

The market mechanism was used to allocate resources for agriculture to a far greater extent than for industry. Although agriculture was included in the national economic plan, the quality of agricultural planning was so poor that central planning of agricultural production was never seriously implemented. Instead, the planners adopted a system of centrally-determined purchase prices as the basic means for determining the relative emphasis to be accorded the various major crops.[25]

The crops hardest hit by the war were industrial crops, notably cotton. The output of cotton in 1949 was only 47 per cent of the 1933 level, while the corresponding figure for grain output was 62 per cent (Perkins, 1966, p. 69). Since the bulk of industrial crop output was exported or used as raw materials by industry, restoration of their production to a normal level was essential to industrial recovery and increased export earnings, both top priority targets of the Peking regime. To stimulate the production of particular crops, purchase prices were raised. As early as 1950, the state set a price for cotton in terms of minimum cotton-grain price ratios to be guaranteed to cotton farmers, supplemented by a list of officially-set absolute prices for major cotton-producing areas (*Ibid.*). These price measures were backed up by improvements in marketing procedures, the most important of which was the introduction of a system of advance purchase contracts. Under this procedure, the state and the cotton farmer agreed in the spring on the price and quantity of cotton to be delivered by the farmer the following fall; the farmer received in advance a certain percentage of the value to be purchased.

Advance purchases and manipulation of the purchase-price structure were said to be quite successful in raising the production and marketing of cotton and other major industrial crops. But such increases were brought about partly at the expense of grain production. By the end of 1952, cotton production had risen to 37 per cent above the 1933 level, while grain production still remained 11 per cent below the 1933 level (Perkins, 1966, p. 69). The slow recovery of grain production, coupled with the rapid rise in demand caused by urbanization and other requirements, resulted in a shortage of

---

25. The system of centrally-determined purchase prices, which was used to stimulate production of particular crops through shifts in relative prices, was never conceived by Communist planners as a long-term approach to raise aggregate agricultural output. The principal measure adopted to raise the level of agricultural production related to increasing unit-area yield.

grain and created serious difficulties when the First FYP began. By the end of 1953 state purchases of grain had run so far behind sales that the resultant deficit could not be continued.

The solution to the grain deficit was the institution of compulsory grain deliveries. This system was initiated in November, 1953, in the form of planned purchases, whereby the peasants were obliged to deliver to the state given quotas of grain at officially-set prices. The delivery quota was supposedly determined on the basis of the requirements of urban deficit areas and the peasants' own needs for food, seed, and fodder. The surplus over the quota could be sold through state commercial organizations at prices slightly higher than state purchase prices. In addition, the supply of grain was rationed to consumers. In 1954, a number of other crops, including cotton, soybean, rapeseed, and peanuts, were also brought under the system of planned purchase.

At about the same time, the state started another system known as unified purchase, which embraced a wide variety of agricultural products not covered by the planned purchase system, including ramie, hemp, sugar cane, tea leaves, draught animals, and live pigs. The unified purchase differed from the planned purchase system chiefly in that the peasants were free to sell the surplus over the quota on the open market and in that the products were not rationed.

By the end of 1954, a large number of agricultural products had been covered under these purchase systems.[26] Compulsory delivery of output quota at low prices may have created disincentive effects among the peasants. Production efficiency may also have been reduced as a result of the administrative and planning difficulties associated with the determination of delivery quotas. But since direct planning of agriculture proved to be unworkable and was abandoned altogether in 1958, the government, for want of better alternatives, had to rely on compulsory purchases.

Prices for agricultural products under quota purchase systems were determined by the state,[27] usually at a level considerably lower than free market prices, so that the latter no longer determined the distribution and production of these products. But there was still a wide variety of commodities, the so-called third-category commodities, which remained outside the quota

26. The list of commodities included in the systems of planned and unified purchases were revised from time to time. In 1959, all commodities produced by non-state industry and agriculture were classified into three categories. The first category consisted of 38 commodities, including food grains, edible vegetable oil, salt, sugar, cured tobacco, raw cotton, cotton yarn, cotton cloth, and a number of petroleum products. The second category covered 293 commodities including ramie, hemp, sugar cane, draft animals, live pigs, eggs, tung oil, tea leaves, fertilizers, insecticides, certain kinds of fruits and a number of aquatic products. Commodities not included in the first and second categories belonged to a third one. Planned or unified purchase was not required for the third category commodities.

27. During the first few years of the system of "planned" and "unified" purchases, state purchase prices were determined primarily by deducting an estimated amount of distribution costs and trade profits from retail prices prevailing in the market. After 1957, the average production cost was used as a basis for setting agricultural purchase prices. See N. R. Chen, 1967, pp. 81, 92.

system. The prices of these commodities were determined basically by supply and demand. Even here, however, production and marketing of these commodities was not left entirely unfettered by the state. Farm subsidiary production is a case in point.

A large number of the third-category commodities, including vegetables, hogs, and poultry, were produced by farmers on their private plots during idle time. When Chinese agriculture was undergoing the process of co-operativization and collectivization, a high proportion of subsidiary production was carried on by individuals, not by cooperatives. Cooperative leadership cadres, in their efforts to achieve production targets, began to encroach on individual independence. The government also adopted administrative measures to control subsidiary production. With the socialization of most commerce in 1955, the free market was in effect closed down. All commodities, including farm subsidiary products, had to be exchanged through state commercial organizations. But, due to the difficulty of setting prices for the great number of commodities produced by subsidiary occupations, the free market was reopened in 1956.

With the advent of the communes, private plots were eliminated and the free market abolished. In consequence, subsidiary production suffered heavily. In view of the important role played by farm subsidiary occupations in providing additional sources of income and employment opportunities to the peasants during their off seasons, the government was forced, in 1959 and 1960, to restore private plots and to reintroduce rural free markets. The subsequent revival of subsidiary production contributed significantly to the recovery of the post-Leap economy.

The present Chinese policy is to tolerate the existence of subsidiary activities but to keep a close watch on them. The government has made every effort to encourage farmers to engage in collective subsidiary production, i.e., subsidiary activities organized and supervised by the commune with income distributed on a collective basis. The growth of individual subsidiary production is discouraged in order to avoid the resurgence of the "capitalist instinct". The trade fairs and other free market activities have been placed under the close supervision of state commercial organizations and market prices are kept under strict control.

Given the fact that agriculture is still the dominant sector of the Chinese economy and that the bulk of the population continues to live in rural areas, farm subsidiary production will undoubtedly remain an important sector of the economy, in which the income-maximizing peasants will continue to be guided in the main by market forces in allocating the limited resources at their disposal.

CHAPTER SEVEN

Conditions of Life and Labor

The size and diversity of the country, the rapidity of change, and the paucity of information combine to make the assessment of Chinese living standards a hazardous enterprise. The only thing that can be done is to compare fragmentary evidence on the physical components of the standard of living with aggregate measures to determine whether the latter can be taken as an adequate representation of trend.

Pre-Communist Conditions

No one will dispute the fact that China was a poor country on the eve of the Communist takeover. While there were countries with even greater poverty, the Chinese worker was hardly affluent. Most of the food energy available to him came from grain and leguminous products, and the rest from vegetables; meat consumption was almost negligible. Buck found that for the years 1924 to 1933, the average daily per capita availability of calories for all of China was 2,365, of which 2,023 came from food grains (Buck et al., 1966, p. 11). Liu and Yeh estimate availability at 1,940 calories from food crops, and 2,130 from all sources, which they defend on the ground that this figure was about the same as for prewar Japan (1965, p. 30). The figure for adults alone would be higher.

For adult workers in Shanghai and Peiping, figures of 2,913 and 2,595, respectively, have been cited (Tao, 1931, p. 15). A recent study arrives at a range of between 2,300 and 3,200 calories *per adult* daily, with an overall average *per person* of 2,200 calories for the years 1931–1937 (Joint Economic

Committee, 1967, p. 262). While the data are not all consistent, they do suggest that workers were by no means on a starvation diet during peacetime years in pre-Communist China. Then, as now, rice was the staple food south of the Yangtze River, supplemented by sweet potatoes (which are considered an inferior food in China). North of the Yangtze, wheat, millet, kaoliang, corn, and soybeans provided the bulk of the diet.

Housing was worse than food. It was said of the countryside that "apart from the frail character of the dwellings of Chinese peasants, it may also be said that they are as a rule unfit for habitation in respect of health and cleanliness" (Tao, 1931, p. 18). Congestion was universal in the cities. The few housing surveys undertaken revealed appalling conditions. Families of handicraft workers and coolies in Peiping averaged 4.16 persons in a room which could range in size from 40 to 120 square feet, but was usually on the lower end of that range. Shanghai cotton mill workers were a bit better off, with 3.29 persons per room, while skilled workers in Peiping averaged 2.49 persons per room.[1] The major Chinese cities seem to have afforded worse housing than Bombay (Tao, 1931, pp. 19–20). The following comment provides a qualitative appraisal of conditions:

The house the Chinese working family occupies has none of its traditional functions, for rest and comfort, for the enjoyment of privacy, for family gatherings or the intimate communion of friends; it has none of these things but a place for sleep and there is often not sufficient space or accommodation even for that ... Squalor, disease, immorality, and crime are but inevitable. It would be strange indeed if a person could live a healthy and moral life under such conditions. (*Ibid.*, p. 21).

Data on the structure of household expenditures in pre-Communist China are shown in Table VII–1. While the surveys on which they are based are of varying quality, it is apparent that food and housing together absorbed about two-thirds of total expenditures. As for clothing, except for the wealthy, cotton cloth was the predominant material used. Average consumption was put at 4.3 meters per capita in 1936 (C. Y. Cheng, 1957, p. 278). Other scattered data place cotton cloth consumption at 3.1 meters for peasants in Hunan Province in 1936, and 6.43 meters for Shanghai workers in 1929–30 (N. R. Chen, 1967, pp. 433 and 440). It has been remarked that the "position of the Shanghai workers in respect to clothing may be supposed as the best in China since the inhabitants of that great metropolis are reputed to have a love for dress" (Tao, 1931, p. 23), but judging from the statistics, Shanghai could not have been a city of sartorial splendor.

Expenditures for fuel depended upon the climate. Peasants used hay, stalk remnants, and brushwood, while a mixture of coal dust and earth was a major source of fuel in the cities. The data in Table VII–1 are misleading in that they reflect a geographical weighting which may not be representative for the country as a whole. Working class families were reported to spend

---

1. A Communist source cites the figures of 14.89 square meters a family and 3.22 square meters per capita for Shanghai in 1929–30. N. R. Chen, 1967, p. 440.

TABLE VII-1
THE STRUCTURE OF HOUSEHOLD EXPENDITURES IN
PRE-COMMUNIST CHINA
(per cent of total expenditures)

Category	Peasants[a]	Shanghai workers 1929–1930[b]	Peasant and worker families[c]
Food	59.9	58.5	57.5
Housing	4.6	8.5	7.5
Clothing	7.1	7.6	7.5
Fuel	10.4	6.1	10.0
Other	18.0	19.3	17.5
	100	100	100

Sources:
[a] C. Y. Cheng, 1957, p. 285. These figures are based on surveys conducted during the period 1922–1934, covering 2,854 households in six provinces.
[b] N. R. Chen, 1967, p. 436.
[c] L. K. Tao, 1931, p. 9. This represents an average of 69 separate surveys, most of them in the cities.

10 per cent of their incomes for fuel, compared with the lower figure for Shanghai (Tao, 1931, p. 24).

The unclassified items in the budget, averaging under 20 per cent, represented a wide variety of expenditures: kitchen utensils, handicraft products, sanitary and medical costs. Educational and cultural expenditures were small.

The Sino-Japanese war, followed by the civil war, encroached heavily upon even these minimal standards. Disruption of farming, physical damage to cities, and the Japanese levies had left the country on the precarious edge of subsistence when the Communists gained power. While there is good reason to believe that the Communist food grain estimate of 108 million metric tons for 1949 was too low (see Buck et al., 1966, p. 48), there is no disagreement that a substantial deterioration of living standards occurred between the onset of the Japanese invasion and the assumption of power by the Communists.

Aggregate Measures of Living Standards Since 1949

As a first approximation to measurement of the welfare of the Chinese people under the Communist regime we will employ two broad measures: national product per capita and food grain available per capita. Neither need be closely related to individual welfare, strictly speaking. National product can increase for considerable periods without affecting living standards, depending upon the magnitude of investment. In the long run, a rising per capita national product is bound to be reflected in improved living standards, but, as the Soviet experience has taught us, it is possible to sacrifice the welfare of an entire generation if there are sufficient political controls. Distribution is a critical matter as far as food availability is concerned, but, generally speaking, in underdeveloped countries good harvests and good eating go hand in hand.

On the statistical side, none of the basic series on which we must rely is firmly established, even for years in which the Communists published data. Nonetheless, they appear to tell a meaningful story which agrees with the accounts of travelers and refugees.

Table VII–2 presents estimated national product and population for the

TABLE VII–2
NET DOMESTIC PRODUCT OF THE CHINESE MAINLAND, 1952–1965

Year	Net domestic product (billions of 1952 yuan)	Population (millions of people)	Net domestic product per capita (1952 yuan)	Index of net domestic product per capita (1952 = 100)
1952	71.4	575	124	100
1953	75.3	588	128	103
1954	79.3	602	132	106
1955	82.3	615	134	108
1956	92.1	630	146	118
1957	95.3	645	148	119
1958	108.0	659	164	132
1959	104.4	669	156	127
1960	95.9	676	142	114
1961	92.2	680	136	110
1962	94.0	687	137	110
1963	98.1	697	141	114
1964	104.2	712	146	118
1965	108.1	728	148	119

Sources:
Net domestic product—T. C. Liu, in Joint Economic Committee, 1967, p. 50. Data for the years 1958–65 are labeled "very crude estimates."
Population: E. F. Jones, 1967b, p. 93.

years 1952 to 1965. The national product figures for the years after 1957 are highly conjectural, and the population estimates are merely demographic extrapolations based upon 1953 figures which are in themselves subject to question. There is a consensus among Western scholars that the national product fell sharply after 1959 and did not regain that level until 1964 or 1965. Because of rapidly increasing population, per capita product in 1965 was probably substantially below that of 1959, perhaps equal to that of 1957, as Table VII–2 indicates.

Estimates of the consumption component of the national product are available only through 1957. These are presented in Table VII–3. The 1957 index of household consumption per capita (1952 = 100) is 109 rather than the figure of 119 shown for per capita net domestic product in Table VII–2.

TABLE VII-3
ESTIMATED PERSONAL CONSUMPTION EXPENDITURES, 1952 TO 1957
(billions of 1952 *yuan*)

	1952	1953	1954	1955	1956	1957
Food	33.08	33.59	36.44	36.35	36.64	38.39
Clothing	7.89	8.79	8.43	7.82	10.40	9.62
Fuel and light	5.11	5.22	5.35	5.46	5.58	5.72
Housing	3.04	3.10	3.16	3.25	3.31	3.40
Miscellaneous	5.43	6.23	7.08	7.89	8.77	9.74
Total	54.55	56.93	60.46	60.77	64.70	66.87

Source:
Estimated by Liu on the basis of the data in Liu and Yeh, 1965, p. 68, and Liu, 1968, pp. 173-174.

The addition of expenditures for communal services does not alter the picture substantially, since the share of such expenditures in the total increased only slightly during the period, from 2.5 to 3.3 per cent (T. C. Liu, 1968, p. 138).

The decline of the portion of total expenditures that went to household consumption from 71.4 per cent in 1952 to 67.0 per cent in 1957 (*Ibid.*, p. 138) was not very drastic, far less than the decline in the ratio during the Soviet First FYP.[2] This may be explained by the fact that the Chinese people were so near the minimum subsistence level in 1952 that there was a very high income elasticity for the basic necessities of life, food in particular, making it difficult for the regime to hold consumption down if this was their purpose. A more plausible assumption is that the Communists wanted to permit consumption to rise, although at a slower rate than the national product, in order to consolidate their power.

The food availability statistics in Table VII-4 tell a somewhat different story from that emerging from the data in Tables VII-2 and VII-3. Per capita availability fluctuated somewhat below the 300 kilograms per annum level from 1949 to 1958, depending upon the harvest. There was a sharp drop, beginning in 1959, to near famine levels. Notwithstanding imports of about six million tons a year, the Chinese had not been able, by 1965, to regain the per capita levels of the mid-1950's due to the inexorable population rise. By this measure, therefore, there may have been some decline in living standards since 1957.

One qualification is in order. During 1952-57, from 16 to 22 per cent of grain output was diverted to such uses as feeding animals, seeding, industrial processing, and exports. This proportion was probably smaller after 1960

---

2. According to Bergson, household consumption outlays fell from 64.7 per cent of Soviet GNP in 1928 to 50.2 per cent in 1937. However, this was offset at least to some extent by an increase in the share of communal services from 5.1 per cent to 14.5 per cent. 1961, pp. 145 and 154.

Table VII-4
FOOD GRAIN AVAILABILITY ON THE CHINESE MAINLAND, 1949–1965

Year	Output (millions of metric tons)	Imports (millions of metric tons)	Total available (millions of metric tons)	Population (millions)	Food grain available per capita (kilograms)	Index of food grain available (1952 = 100)
1949	150		150	545	275	93
1952	170		170	575	296	100
1953	166		166	588	282	95
1954	170		170	602	282	95
1955	185		185	615	301	102
1956	180		180	630	286	97
1957	185		185	645	287	97
1958	204		204	659	310	105
1959	170		170	669	254	86
1960	160		160	676	237	80
1961	170	6.2	176	680	259	88
1962	180	5.3	185	687	269	91
1963	185	5.7	191	697	274	93
1964	195	6.8	202	712	284	96
1965	200	5.7	206	728	283	96

Sources:
Grain output: Joint Economic Committee, 1967, p. 93.
Grain imports: *Ibid.,* p. 601. Exports, amounting to a maximum of one million tons during these years, are not deducted from the availability data, since the figures are not available for calendar years.

due to scarcities, so that grain available for human consumption may not have declined by as much as Table VII–3 indicates.

Urban Living Standards, 1952–1956

The first step in the process of disaggregating these broad measures of welfare is to separate the urban and rural sectors of the economy. There is by no means a clean cut distinction between the two, for geographical variations may have been more important than size of community. However, food is more easily procured in rural areas and manufactured goods in cities, as a rule. There are differences in housing requirements, services, and other elements in the family budget which make this a meaningful distinction.

Wage and price estimates are shown in Table VII–5. The data for the years 1950 to 1957 are derived from official Chinese sources, while those for subsequent years are Western estimates. They purport to show that real wages rose steadily to 1957 and then fell precipitately with the onset of the Great Leap Forward. If the data are to be credited, the 1965 level of real wages was still below that of 1957, despite subsequent recovery. Compare Table VII–5 with Tables VII–2 to VII–4: per capita product peaked in 1958 at

TABLE VII-5
MONEY AND REAL EARNINGS OF CHINESE WORKERS
AND EMPLOYEES, 1949–1965
(*Yuan* per annum)

Year	Money wages	Cost of living index (1952 = 100)	Wages in 1952 prices	Index of real wages (1952 = 100)
1949	262[a]			
1950	322[a]	88.6[d]	363	81
1951	379[a]	99.1[d]	382	86
1952	446[b]	100[e]	446	100
1953	496[b]	105.6[e]	470	105
1954	519[b]	106.9[e]	486	109
1955	534[b]	107.3[e]	498	112
1956	610[b]	107.1[e]	570	128
1957	637[b]	109.2[e]	583	131
1958	551[a]	108.3[d]	509	114
1959	531[a]	108.3[f]	490	110
1965	583[c]	108[g]	540	121

Sources:
[a] CIA, 1960, p. 3.
[b] N. R. Chen, 1967, p. 492 (official Communist data).
[c] See text.
[d] N. R. Chen, *ibid.,* p. 409. This is an index of retail prices in eight large cities.
[e] *Ibid.* This is an index of living costs for workers and employees in twelve cities.
[f] Dwight H. Perkins, 1966, p. 156.
[g] Based upon U.S. Consulate General, September 19, 1966. This source, from Communist China, states that for years since 1959, the rise and fall of the price index did not exceed 1 per cent a year, and that prices for basic necessities changed by less than 0.1 per cent a year.

132 per cent of the 1952 level; grain availability also peaked in 1958, but only at 105 per cent of 1952; while urban real wages reached their maximum in 1957, when they were 131 per cent of 1952.

The figures for 1965 are highly speculative, and it is difficult even to judge their approximate validity. Japanese industrial experts visiting machine tool plants in 1965 and 1966 were told uniformly that workers' earnings averaged 60 *yuan* a month (JPRS, July 17, 1967). A British visitor commented that at every factory she was told that 60 *yuan* a month was the average wage, but she felt that 45 to 50 *yuan* was nearer the mark (MacDougall, Dec. 23, 1965). The reasonableness of the latter figures may be argued as follows: the 60 *yuan* level was for metal processing plants (Peking Machine Tool Plants No. 1 and 2, Mukden Machine Tool Plant No. 1), where wages would be expected to exceed the average level. In 1956, average annual earnings of all workers and employees were 610 *yuan*. In the same year, workers and employees in metal processing enterprises operating within the state plan earned 751 *yuan* (N. R. Chen, 1967, p. 494). On the assumptions a) that the ratio between national average and metal processing wages remained constant, and b) that the plants visited by the Japanese were important enough to be operating within the state plan (which seems highly likely), the 720 *yuan* annual figure for 1965 (60 *yuan* a month) would be reduced to 583 *yuan* per annum, which is the figure shown in Table VII–5 (For substantiation, see *Far Eastern Economic Review*, Sept. 1, 1966, p. 417, and Sept. 8, 1966, p. 441).

The basic question to be answered is this: could real wages have risen as rapidly up to 1957 as Table VII–5 indicates, given the tremendous investment program that was being carried on concurrently?

From 1952 to 1957, the ratio of so-called workers and employees to total nonagricultural employment rose from 43 per cent to 62 per cent (Emerson, 1965, pp. 128–29). Workers and employees were found mainly in the modern, state sector, while employees not so classified worked in handicraft and traditional trades. Reported annual earnings in the private sector, which consisted mainly of the latter groups, rose a bit less than those in the state sector from 1952 to 1956, 35 per cent as against 39 per cent (Schran, 1961, p. 277). Thus, the wage data in Table VII–5 would tend to overstate the degree of income improvement for the urban population as a whole, although the amount of the overstatement is not great, and tends to diminish over the period as the worker and employee group become more representative of the entire urban labor force. The problem can be posed symbolically as follows:

(1) $$\Delta Y = \Delta C_1 + \Delta C_2 + \Delta C_3 + \Delta T + \Delta S$$

where ΔY is the increase in spendable income, ΔC_1 is the increase in food consumption, ΔC_2 the increase in the consumption of non-food commodities, ΔC_3 the increase in the consumption of services, ΔT the increase in taxes, and ΔS the increase in savings. The variables can be expressed either in constant price valuations or in kind. For expositional purposes, we will tackle the problem in the following order:

(2) $$\Delta Y - \Delta T - \Delta S = \Delta C_1 + \Delta C_2 + \Delta C_3$$

We will also restrict ourselves initially, to the period 1952 to 1956, for which the data are relatively good.

First with respect to ΔY itself, Liu and Yeh argue: "It is possible, of course, that the official indexes were constructed largely on the basis of official prices charged in state-operated stores. Prices charged in private transactions may have been quite different. There is also the possibility that there were clandestine transactions, even in state-operated stores. It is doubtful that all these unofficial prices have been given proper weight" (1965, pp. 106–07). In this view, part of the apparent increase in urban purchasing power was siphoned off either to the countryside or to industrial enterprises. Perkins concurs: "These indexes ... do not completely reflect the true extent of price increases during periods of greatest inflationary pressure, 1953, 1956 and 1960–1962. This was the result of the inability of the regime to collect adequate nationwide price data for any but a few dozen major consumer items. ... The remaining commodities, the ones that were not generally included in the indexes, not only were not subject to such rigid price control, but had, in addition, to soak up excess purchasing power that normally would have been soaked up by major commodities. These excluded commodities made up approximately 30 per cent of all retail sales and were sold mostly on the free markets" (Perkins, 1966, pp. 156–57). He estimates that the resultant downward bias in the price index was 3 or 4 per cent at a maximum. If we accept the lower figure, the cost of living figure for 1956 becomes 110.3 and the "true" real wage index for the same year falls to 124, reducing the living standard increase to be explained.

Turning next to ΔT, we find that this was not a significant factor. No personal income tax was imposed, while increased excise taxes would presumably have been reflected in the consumer price index. As for ΔS, a number of domestic bond issues were floated during the period, and subscriptions may have been compulsory. However, the amounts taken up by urban employees showed no clear tendency to increase (Ecklund, 1966, p. 87). The one item that did show a consistent upward trend, urban savings bank deposits, rose from 11 per cent of the urban wage bill in 1952 to 16 per cent in 1956 (Perkins, 1966, pp.161 and 170). To the extent that these deposits were subject to future withdrawal, they represent a shift of income to the future. Comparing 1956 and 1952, however, the savings increases implied a 5 per cent reduction in spendable earnings, bringing the wage index for 1956 down to 119.

On the left hand side of our equation, then, there is a 19 per cent increase in spendable real income to be reconciled with increased consumption of goods and services.[3] If this figure, representing 4.75 per cent a year, reflected reality, it would have been no mean achievement, considering the pace of industrialization.

There are several family expenditure budgets available for years *circa* 1956, which appear in Table VII–6. These are Communist data. What purports to be a national sample and the Shanghai data are in close agreement, but the budget for Kirin, a province of Manchuria, differs substantially from the others. We will accept the national-Shanghai data as representative for urban areas in general.

Food absorbed more than half of total expenditures.[4] The items may be divided into staples and other foodstuffs. A rationing system for food grains was established in 1955, but informal rationing began several years earlier (C. Y. Cheng, 1957, pp. 308–11). The output of grain was estimated to have risen from 170 million tons in 1952 to between 175 and 180 million tons in 1956 (Joint Economic Committee, 1967, p. 92). What this meant to the urban wage-earning family depended upon two things: the additional number of

---

3. There is an implicit assumption that real income, measured in this way, moved proportionally with real income per capita. If average family size increased during the period, or if money wages diverged from income by virtue of a change in the ratio of wage earners to the self-employed, which may have occurred because of migration to the cities, the assumption would not be valid.

4. The 1956 proportion spent on food is less than that reported for pre-Communist China. See Table VII–1.

TABLE VII-6
THE STRUCTURE OF FAMILY EXPENDITURES OF
WORKERS AND EMPLOYEES
(per cent of total)

Category	National sample, 1955[a]	Workers in Shanghai, 1956[b]	Kirin Province 1956[c]
Food	55	52.53	44
Staples	n.a.	15.59	21
Other	n.a.	36.94	23
Nonfood products	25	24.44	41
Clothing	12	11.88	18
Other	13	12.56	23
Services	20	23.03	15
Rent	n.a.	1.94	4
Other	n.a.	21.09	11
Total	100	100	100

Sources:
[a] N. R. Chen, 1967, p. 435.
[b] *Ibid.*, p. 436.
[c] *Ibid.*, p. 432.

mouths to be fed, and the distribution of grain between country and city. On a national per capita basis, these estimates may be translated into a reduction from 296 kilograms in 1952 to 286 kilograms in 1956 (Table VII–3). The ratio of urban to total population increased somewhat from 1952 to 1956 (Hou, 1968, p. 342), so that urban per capita grain availability would have declined even more unless counterbalanced by a more favorable distribution from the urban point of view. It seems conservative to assume that urban grain consumption per capita remained stable during the period 1952–1956.[5] We must look elsewhere, therefore, for the increase in consumption.

The predominant portion of expenditures for food went for non-staple items, according to the Shanghai family budget study of Table VII–5. This does not mean a similar ratio in terms of nutritive value. From 80 to 90 per cent of the caloric intake was provided by cereals. The disparity reflects the government policy of keeping staple prices low and permitting the sale of supplemental foods at a much higher price level.

There are only a few fragments of information on changes in consumption of foodstuffs other than grain during the period in question: (a) vegetable consumption per capita in Peking was said to have risen 47 per cent from 1952 to 1956, and pork 26 per cent (N. R. Chen, 1967, p. 437); (b) per capita meat consumption in Kansu Province was said to have risen by 50 per cent

---

5. This is the conclusion reached by T. C. Liu in Joint Economic Committee, 1967, p. 59. C. Y. Cheng (*ibid.*, p. 315) argues, on the basis of scattered evidence, that grain consumption fell during the period.

from 1954 to 1956, and that of edible oil by 58 per cent (*Ibid.*, p. 439). These figures purport to be representative of the country as a whole, but they are not compatible with the following Western assessment of output: "There was no appreciable increase in the production of other major food crops during the 1952–57 period. Some increases occurred in the production of fruits and vegetables on private plots but vegetable oilseeds, which provide a substantial source of fats in the diet, remained static; peanut production increased slightly but soybeans declined slightly" (Joint Economic Committee, 1967, p. 258).

The one piece of information from official Chinese sources that seems at all credible is the claim that per capita consumption of subsidiary foods, measured in constant *yuan*, rose by 18 per cent from 1952 to 1956.[6] On this basis, we estimate that living standards rose by 6.7 per cent from 1952 to 1956, on account of the increased consumption of non-grain foods.[7]

This conclusion, once again, is conservative when viewed against the production estimate cited immediately above. One must also be aware of the welfare implications implicit in the calculation. As already noted, supplementary foods provided the Chinese people with a much smaller nutritive value than their place in the family budget would indicate. Since grain was rationed, the consumer was obliged to supplement his diet with higher priced foods, although they did, of course, provide variety. The actual nutritive increment to the diet provided by non-cereals was considerably less, therefore, than the family budget percentages indicate; the two would presumably coincide only if all foodstuffs were unrationed and sold on the basis of free market prices. In effect, we overstate the welfare gain by the procedure adopted.

If, nevertheless, we accept 6.7 per cent as being the standard of living increase on account of supplementary foods, we are left with a 12.3 per cent increase still to be explained by other items.

The next major category is clothing, which ranked second only to grain as an article of consumption in China, although its weight in the family expenditure budget was only 12 per cent. This was again due to relatively low prices, which forced rationing beginning in 1954. The various pieces of information about clothing availability, much of it conflicting, are shown in

6. Nai-Ruenn Chen, 1967, p. 433. The same source shows a slight *decline* in the consumption of food grains during the same period. There are also data showing the consumption of "other" food products: cigarettes, liquor, tea, and restaurant expenditures. These are small items, however, and we will regard the "subsidiary" food category as representative of non-cereal food consumption.

7. The calculation is as follows: the portion of the family budget devoted to non-staple foods in 1956 (Shanghai sample) was 37 per cent; the consumption of such foods rose by 18 per cent; the product of the two is 6.7 per cent. This figure must be recognized as being highly speculative on several counts: 1) the Kirin Province family budget for 1956 shows a weight of only 23 per cent for non-staple foods; we are assuming that the Shanghai budget is more representative; 2) it is assumed that expenditures in current prices for non-staple foods were deflated by the Chinese statisticians by a price index specific to this item, rather than by a general price index; 3) the most dubious assumption is that the Shanghai experience is representative for all of China.

Table VII-7. There was a large increase in textile exports during the period, but the absolute amount in relation to domestic consumption was small.

There is little question that the consumption of cotton cloth rose rapidly from 1952 to 1956. Perkins notes that "the income elasticity of demand for cotton cloth was high. One source states that it was the first commodity upon which increases in consumer income were spent" (Perkins, 1966, pp. 187–89). But there appears to be no way of reconciling the scraps of official data. We will cut the Gordian knot by using the figure 43 per cent, on the quite arbitrary assumption that peasants and workers fared equally well in this respect.[8] On this basis, the addition to the standard of living attributable to clothing comes to 5.2 per cent (43 per cent × 12 per cent). If the lower Liu-Yeh estimate had been used, the effect would have been to reduce the four year increase by 1.4 percentage points.

TABLE VII-7
DATA RELATIVE TO CLOTHING AVAILABILITY, 1952 TO 1956

	Per cent increase, 1952 to 1956
Official Chinese data	
Consumption per capita of machine-made cloth, entire country[a]	61–66
Per capita expenditures on clothing of workers and employees in Kirin Province[b]	−14
Amount of cotton cloth consumed by peasants[c]	43
Cotton cloth production by factories[d]	52
Per capita expenditures on clothing of peasants in Hunan Province[e]	67
Western estimates	
Clothing consumption at consumers' prices, entire country[f]	32
Retail sales of cotton cloth, entire country[g]	81

Sources:
[a] N. R. Chen, 1967, p. 436.
[b] *Ibid.*, p. 433.
[c] *Ibid.*, p. 438.
[d] *Ibid.*, p. 197.
[e] Liu and Yeh, 1965, p. 261.
[f] Perkins, 1966, p. 188.
[g] N. R. Chen, *ibid.*, p. 433.

There are scattered data available for consumption of the remaining non-food consumer goods, but no satisfactory aggregate statistics. The best summary of the situation is provided by Liu and Yeh: "From 1952 to 1956 the Communists published some aggregate data on daily consumption items including china and earthenware; consumers' metal products; leather and fur products; glass products; cultural, educational, and technical products; and

---

8. This does not mean, however, that the two groups received the same absolute rations of cloth. In 1955, workers and employees consumed 11.37 metres of cotton cloth per person, compared with 8.8 metres for peasants. N. R. Chen, 1967, p. 438.

an unnamed 'others' category which varied in size from one-fifth to one-third of the total. The gross value of output of this aggregate group of consumers' goods increased 44 per cent from 3.7 billion *yuan* in 1952 to 5.3 billion *yuan* in 1956" (1965, p. 61). For want of a better figure, we shall accept 44 per cent as representative of the increased availability of this range of goods,[9] thus yielding a standard of living increment of 5.7 per cent.

Thus far, we have estimated that $\Delta C_1 = 6.7$ per cent, while $\Delta C_2 = 10.9$ per cent, compared with a total spendable wage increase of 19 per cent to be explained. If these figures are sound, the 20-23 per cent of the family budget devoted to rent and other services remained almost unchanged during the period. It remains to determine whether this is a reasonable conclusion.

We run once again into the pricing problem, this time with respect to housing. According to Table VII-6, expenditures for rent absorbed only 1.94 per cent of family budgets because of the Communist low-rent policy. If this figure is taken at face value, changes in housing standards become almost irrelevant to the measurement of standards of living, which is a patent absurdity. On the basis of one estimate, per capita living area in the cities declined by 25 per cent from 1952 to 1956, apart from an additional decline in quality (Kang Chao, 1968b, Table 5). We shall allow for this factor by using an alternative budget; for the present, we simply note that using the budgets in Table VII-6, living standards would have declined by 0.5 per cent on this account (25 per cent × 1.94 per cent), raising the unexplained residue to 1.9 per cent.

The Kirin Province budget for workers and employees breaks down per capita noncommodity expenditures other than rent, water, and electricity as follows (N. R. Chen, 1967, p. 433. Data are in 1952 *yuan*).

	1952	1956	Per cent increase, 1952–1956
Service payments	3.95	8.46	114
Culture and recreation	1.84	1.59	−14
Other	6.56	8.68	32
Total	12.35	18.73	52

Since there is no indication of what "service payments" and "others" include, these figures cannot be tested. If they are representative of all urban areas, however, they are difficult to reconcile with the fact that employment in the industries which presumably provided the services in question declined from 1952 to 1956. The largest employer of service labor, trade, showed a decline from 8.45 million to seven million employees; the next largest, traditional transport, from 3.5 million to 2.5 million employees. Modern transportation and finance increased their employment, but in absolute

---

9. Liu and Yeh regard this as too high an estimate. 1965, p. 63.

numbers, they were outweighed by the first two sectors (Emerson, 1965, p. 128). Unless the productivity of the service occupations grew rapidly enough to compensate for the decline in employment—and these tend to be the industries in which productivity increases slowly—the implication is that the volume of services available to the Chinese consumer declined rather than increased from 1952 to 1956.[10]

If we make the assumption that the availability of the services other than rent remain unchanged during the period, we end with the conclusion that urban real income, measured in this way, may have risen by 17 per cent, or 4.25 per cent a year, exclusive of additional services provided free by the state. Before accepting this figure, however, we must consider the implications of the family budget structure that we used in our estimates. As already indicated, in the absence of market prices which are related to opportunity costs, family expenditure data may not reflect the welfare considerations implicit in our use of the term "standard of living." This is particularly true where vital commodities are subject to rationing. Consumers may be obliged to spend their money on costly non-necessities for lack of availability of what they really want. Indeed, this may be regarded as a form of taxation when a government wants to close an inflationary gap without directly reducing incomes.

How the Chinese consumer might have preferred to allocate his earnings if a market price system had prevailed is best indicated by pre-Communist family budget studies, the results of several of which are shown in Table VII-8. The Peiping data differ substantially from the others, perhaps because of differences in the area, the sample, or the budget classifications. The others are in substantial agreement with respect to food, but differ somewhat in the remaining categories. We will use the data emanating from a Communist source since it appears in the same set of tables as the 1956 statistics on which we relied above. Living standard changes from 1952 to 1956, with these budget weights, would have been as follows:

Staple foods	0
Subsidiary foods (18% × 29.6%)	+5.5%
Clothing (43% × 7.6%)	+3.3%
Other non-food products (44% × 9.8%)	+4.3%
Housing (−25% × 8.5%)	−2.1%
Total	+11.0%

This calculation puts the rate of improvement of urban living standards at 2.75 per cent a year from 1952 to 1956, compared with the 4.25 per cent figure obtained above. We would not want to assert that the lower figure represented the true improvement in living standards enjoyed by Chinese

---

10. Employment in state education, medicine and public health, and cultural affairs did increase, but since they were financed by the government, they would not have been reflected in family expenditures.

TABLE VII-8
FAMILY EXPENDITURES IN PRE-COMMUNIST CHINA
(per cent of total)

Category	Shanghai workers 1927–1928[a]	Shanghai workers 1929–1930[b]	Peiping workers 1926[c]	69 budget studies 1924–1930[d]
Food	56.0	58.5	34.5	57.5
Staples	29.8	28.9	17.5	
Other	26.2	29.6	17.0	
Nonfood products	23.2	17.4	22.9	
Clothing	9.4	7.6	8.4	7.5
Other	13.8	9.8	14.5	
Services	20.8	24.1	42.6	
Rent	6.4	8.5	11.5	7.5
Other	14.4	15.6	31.1	
Total	100	100	100	

Sources:
[a] Yang Tao, 1931. Fuel and light are included in "other" nonfood products, which also includes cigarettes, toilette, furniture and utensils, and articles of religious worship.
[b] N. R. Chen, 1967, p. 436. These data are from a Communist source.
[c] Gamble, 1933. Food expenditures are divided into "staples" and "others" on the basis of data for the median income group. "Other nonfood products" include health, house equipment, religious objects, and incidentals.
[d] Tao, 1931.

urban dwellers during the years in question. There are many arbitrary assumptions embedded in our estimates, and even such data as we were able to marshall do not have a high degree of reliability.

But the calculation does throw serious doubt on the reasonableness of the conclusions that can be drawn from the Communist real wage data shown in Table VII-5, even as adjusted downward for bias in the price index. Scepticism was expressed earlier about the possibility of combining an extremely heavy investment program with a 4.75 per cent per annum improvement in urban living standards.[11] The dichotomy between wages and living standards, resulting in the lower figure of 2.75 per cent, provides one clue to an appropriate discount of the wage data. The assumptions that we have made tend to yield a higher rate of increase than might have obtained were we stricter. Some discounting may be in order on this account.[12]

Yet when all is said and done, even the maintenance of constant living standards during this period would have been no mean achievement. The

11. Robert Dernberger, in a letter to the authors, has argued that preferential treatment to urban workers, a relatively small group in the population, may have made this combination possible. However, whatever evidence we have been able to secure points to the conclusion stated in the text.
12. Ta-Chung Liu has estimated that the average annual rate of per capita consumption increase, measured in constant 1952 prices, for both urban and rural areas, was 1.85 per cent from 1952 to 1957. See Table VII-3.

Communists appear to have been able to combine rapid growth with tolerable living standards for the population.

Urban Living Standards Since 1956

According to the data in Table VII–5, average real wages rose slightly in 1957, then declined sharply in 1958 and 1959. The 1958 drop, at least, was a statistical rather than a real phenomenon. During the year there was a sharp increase in newly hired unskilled workers accompanying the Great Leap, which tended to reduce the average wage, though an individual worker who retained his old job would not have suffered a decline in income. In 1959, however, the Chinese had the first of three bad crops and by 1960, if not earlier, a substantial deterioration of living standards had set in. This period has been described in the following terms:

> The gap between total production and the subsistence requirements of people, which had opened after 1949 with the coming of peace, now closed. Two poor harvests in 1959 and 1960 resulted in a dangerous cut in food rations. The winter of 1960–61 marked the lowest point in food rations under the Chinese Communists. It is the only period when visitors to Communist China detected signs of widespread malnutrition. It was a time of greatly weakened resistance to disease, of grave discontent among the population which affected even the army, and probably of numerous outright deaths from starvation. (Ashbrook, 1967, pp. 31–32).

The winter of 1960–61 marked the low point in the economic fortunes of the Chinese workers; living conditions improved gradually thereafter. Per capita food availability increased with improving harvests, though less rapidly than farm output because of the growing population. How much food the average Chinese consumer had in 1965 and 1966 is a question on which there is considerable disagreement, in part because of differences of opinion regarding the size of the population.[13]

In 1967, staple grains and edible oil were still being rationed. Such subsidiary foods as meat, fish, poultry, and vegetables were obtainable seasonally without restriction, but at prices considerably higher than the controlled ration prices. Since 1963, foreign visitors to China have found no evidence of food shortages in the cities.[14] How the rations compare with 1958 it is impossible to say.

The textile situation appears less favorable. Chinese statistics indicate that the annual consumption of cotton cloth per worker in Shanghai was 14 metres in 1956 (N. R. Chen, 1967, p. 440). The yearly cloth ration for workers, officials, and students in Peking was reported to have been 4.3 metres in 1964, with housewives and the self-employed receiving only about half as much (*Far Eastern Economic Review*, Dec. 31, 1964). The average ration in Shanghai

---

[13]. See, for example, the discussion between Klatt and Perkins in *The China Quarterly*, July–September, 1967.

[14]. See the *Far Eastern Economic Review*, Aug. 15, 1963, p. 370; Feb. 18, 1965, p. 277; Aug. 19, 1965, p. 317; and Feb. 23, 1967, p. 266.

was said to have been 6.7 metres per person in 1966, compared with four metres in 1965 and three metres in 1964 (*Ibid.*, Dec. 16, 1965). It may be that the latter figures are properly compared with per capita, rather than per worker consumption in 1956; this figure was 8.36 metres per capita for the entire country (N. R. Chen, 1967, p. 436). Even without adjusting for the fact that urban rations exceed rural rations, there appears to have been a deterioration in clothing standards between 1956 and 1966.

The housing situation may be the worst of all. A visitor to Peking in 1964 gave the following account of construction activities:

Recently building has started again here and there, after having come to a complete halt in 1960. One of the saddest sights of Peking is a vast new housing complex half-built, over part of which still towers a giant crane apparently rusting in its bed. There are many abandoned construction sites to be seen around Peking, as in other Chinese cities, waiting for a new wave of prosperity to set them going again. But there are also isolated cases of new buildings now going up with work going on, particularly in Peking. (*Far Eastern Economic Review*, July 16, 1964, p. 117).

An article in the Chinese press, appearing about the same time, complained that a factory had authorized an investment of 2.4 million *yuan* in 1959 to build 30,000 square metres of housing for 1,153 families, but by 1963, with 3.2 million *yuan* already expended, only 26,000 square metres, housing 215 families, had been completed. The rooms were too large; big windows made it difficult for workers on the night shift to sleep; and rooms on the north were too cold (U.S. Consulate, Hong Kong, Feb. 25, 1965).

A Western journalist who visited Shenyang, an industrial city in Manchuria, in 1965, described the home of the head of a workshop in a machinery repair plant, who earned 100 *yuan* a month (twice the average wage), as follows: the family and four children lived in a single room 13 feet by 8 feet in size, a good part of which was taken up by two large beds. Rent cost only 3.22 *yuan* a month, while 10 *yuan* a month went for gas and electricity. As crowded as this dwelling was, it was still far better than the slum housing in the back streets of Peking, or in Shenyang in general (MacDougall, Dec. 23, 1965).

The question seems to be not whether housing conditions have improved since 1956, but whether housing construction has kept pace with the growing population. For several years after the Great Leap there appears to have been a construction stoppage, and there may have been another during the Cultural Revolution of 1966–67. The Chinese have not been making any boasts about their housing achievements, and it is a fair surmise that, as in the case in the Soviet Union, housing will be the component of consumer living standards which will improve most slowly.[15]

A wide range of other consumer goods is available in the open market, but prices are high. For example, in 1965 a bicycle cost 125 *yuan* and up,

---

15. It has been estimated for Shanghai that per capita living area fell from 3.7 square metres in 1956 to between 2.15 and 2.33 square metres in 1965, with perhaps even sharper declines in some of the industrial cities that were expanding more rapidly. Howe, 1968, p. 73.

the cheapest steel-encased wristwatch, 90 *yuan*, and a portable one wave-band transistor radio, 50 to 70 *yuan* (*Far Eastern Economic Review*, Aug. 19, 1965, and Dec. 23, 1965). Fourteen-inch television sets were being sold in Shanghai for 430 *yuan*, but in view of the fact that average worker income was about 50 *yuan* a month, these were sold mainly to institutions.

What can be said, in summary, is that the decade following the completion of the First FYP was not marked by any overall improvement in the standard of life enjoyed by the Chinese urban dweller. There is a possibility that food was in shorter supply in 1966 than in 1956, and it is quite unlikely that there was any improvement.[16] Clothing would appear to have been in shorter supply than in 1956, while housing did not improve, and may have actually deteriorated. In sharp distinction to the First FYP period, 1952 to 1957, when China managed to combine rapid industrial growth with some improvement in consumption standards, the decade 1958–1967 was characterized by industrial stagnation and, if anything, a retrogression in living standards. The decline of investment in heavy industry does not seem to have been marked by a corresponding increase in investment in consumer goods industries and in agriculture. What seems to have occurred was a general decline in economic activity, leading to a smaller absolute amount of investment resources available.

Rural Living Standards

Difficult as it is to put some quantitative limits on urban living standards, it is far more difficult to be precise about the lot of the 80 per cent of the Chinese people who live in villages and who derive their income from farm work and subsidiary occupations. The few Chinese statistics available, shown in Table VII–9, point to a differential in living standards adverse to the peasant. However, neither income nor expenditures per se can portray adequately the relative living standards in country and city because of differences in prices and in consumption patterns, and because of the ability of peasants to supplement purchased food and clothing with their own production.

At the time of the Hundred Flowers episode many complaints from peasants, to the effect that industrial workers were enjoying higher standards of life, came into the open. The government replied that from 1952 to 1957, the prices of industrial goods in relation to farm prices had fallen 14 per cent, but published data indicated that the workers still enjoyed a considerable edge (Huenemann, 1966, p. 48).

16. Dwight Perkins, who tends to be on the high side of the grain output estimate spectrum, thought that 200 million plus tons of grain was a reasonable estimate for 1965, and that 1966 probably saw some improvement. Perkins, 1967. Even if one were to allow for a 5 per cent increase in 1966 over 1965, and for six million tons of imports, with a resultant total of 216 million tons, food grain availability per capita in 1966 would have been 290 kilograms compared with 286 kilograms in 1956 and 310 kilograms in the bumper crop year of 1958. (The 1966 figure is based upon an estimated population total of 744 million for that year).

TABLE VII-9
RELATIVE INCOME AND EXPENDITURES OF WORKERS AND
EMPLOYEES VERSUS PEASANTS IN COMMUNIST CHINA

	Urban-workers and employees	Rural-peasants	Percentage differential, workers and employees over peasants
National average, 1955			
Income, current yuan	102	94[a]	8.5
Expenditures, current yuan	97	89[a]	9.0
Per capita consumption of food grains 1956–57 (kilograms)	282.4	258.9	9.1
Per capita consumption of cotton cloth, 1955, meters	11.4	8.8	29.5

Source:
N. R. Chen, 1967, pp. 430, 437, and 438.
Note:
[a] Excludes the consumption of handicraft products made by peasants for own use.

The amount of grain consumed by peasants appears to have remained stable, at best, during the years 1952–1957 (Schran, 1961, p. 241). The consumption of other foods and textiles probably increased. The greatest relative gain made by the peasants was probably in the increased availability of manufactured goods at favorable prices (*Ibid.*, p. 141).

The aftermath of the Great Leap Forward probably marked some leveling of urban-rural living standards if only because the peasants were at the source of scarce food, and because of the shift of emphasis from industry to agriculture. It was asserted that in 1964, a peasant could buy 50 per cent more industrial goods for the same amount of farm produce as in 1950.[17] Peasant income still seems to be substantially below that of workers (*Far Eastern Economic Review*, Nov. 3, 1966, p. 263), but this is not necessarily translated into lower living standards.

Housing is a case in point. Neither in the city nor in the country is a substantial portion of the family budget devoted to rent. The older housing in the cities may not be much better, if indeed it is at all better, than peasant housing;[18] but, on the average, urban housing standards are probably better than rural standards. On the other hand, the deterioration of the urban housing stock from 1957 to 1967, and the lack of space in the cities, contrasted with the ability of peasants to maintain and add to their very

---

17. U.S. Consulate General, Hong Kong, *Selections from China Mainland Publications*, No. 542, September 19, 1966.
18. The following description of workers' housing in Peking prior to the Communist regime tends to bear out this assertion: "A cheaper type of wall is made of adobe brick faced with a mixture of earth and lime. The roofs of the cheaper houses are made of an earth and lime mixture, or sometimes simply of packed mud. When these become old, they are apt to leak during the summer rainy season. . . . Most of the floors are of dirt or tile." Gamble, 1933, pp. 121–122.

simple dwellings to accomodate larger families, may have reduced the spread between country and city housing.[19]

How well the Chinese peasant is doing at the present time is a matter on which there is profound disagreement among the experts. There are two critical variables, food grain output and population, and much depends upon which of conflicting estimates can be regarded as reasonable. Owen Dawson, the former U.S. agricultural attache in China, estimated a total grain output of 193 to 200 million tons for 1965, whereas the Agricultural Officer of the U.S. Consulate General in Hong Kong estimated only 180 million tons of output for the same year. Dwight Perkins and Alexander Eckstein supported the higher estimate, which would imply a slight decline in food availability per capita since 1957 if one accepts the population figures in Table VII–2. Perkins argues that the increase in the use of chemical fertilizer should have raised grain output by 16 to 20 million tons since 1957 (Perkins, 1967, p. 38), while Eckstein feels that "by 1965 the per capita food position does not appear to be perceptibly worse than in 1957" (Joint Economic Committee, 1967, p. 179). Ta-Chung Liu, on the other hand, argues for the lower estimate; it is his position that the higher estimate, after adjustment for the difference in age composition of the population, would suggest higher per capita grain consumption in 1965 than in 1957. If this were the case, the Chinese would scarcely be willing to spend 30 per cent of their foreign exchange earnings to import grain (*Ibid.*, 1967, p. 41). Werner Klatt, a British expert, has taken the position that grain output in 1965 was substantially below the Dawson estimate, but that food rations were adequate because the population in that year was not more than 700 million, and possibly less, rather than the 728 million figure that we have used (*China Quarterly*, July–Sept. 1967, p. 151).

Average per capita calorie intake in 1958, which most people agree to have been the best crop year in recent Chinese history, has been estimated at 2,200 (Larson, 1967, p. 265), a level which "under Chinese conditions . . . was likely to provide the population with ample food for heavy work. At that level the population would remain in reasonably good physical condition. Below that intake there was bound to be loss of body weight, retarded growth among children, and reduced output among working men and women (Kaye, 1961, p. 126). Assuming that caloric availability is proportional to grain availability, the daily per capita caloric intake in 1965 would have been 2,000 calories (based on the data in Table VII–4), using the higher of the crop estimates.[20] If one were to accept the lower grain output figure of 180 million tons for 1965, and add imports, the daily calories available that year would have been only 1,800 per capita, which is not far from a starvation diet.

19. A visitor to Kwangtung Province in 1966 had the impression that village housing was better than city housing in that area. *Far Eastern Economic Review,* Feb. 23, 1967, p. 267.
20. This is also the estimate of Marion R. Larson, 1967, p. 267. There is some evidence of an improvement in the quality of the diet, which may have meant a higher ratio of calories to grain in 1965 than in 1958, and therefore a figure in excess of 2,000 calories for 1965. Dr. Werner Klatt argues for a level of 2,100 calories for 1966–67, when the grain output may not have been as high as in 1965. 1967, p. 154.

Analogy with India may help clarify the significance of these figures. India enjoyed a bumper crop in 1964 (the 1965 crop was disastrous). Together with substantial imports, a total of 99.5 million tons of food grain were available in 1964 (FAO, 1966b, p. 81). With a population of 475.4 million in the same year, the per capita availability of food grains was 209 kilograms, substantially below even the low Chinese estimate for 1965. However, the Food and Agricultural Organization has estimated that the per capita caloric intake in India for 1964–65 was 2,110 per day (*Ibid.*, p. 425). A portion of the discrepancy may be explained by taking into account differences in the pattern of food consumption between the two countries.[21] Cereal grains and potatoes (the latter a very small item) provided 69 per cent of the calories in the Indian diet in 1965, but 80 per cent of the Chinese diet (FAO, 1966b, p. 425, and Larson, 1967, p. 267). There is still a considerable discrepancy, however; if the calorie/grain ratio were the same for China as for India, the Chinese caloric availability would have been 2,500 rather than 2,000.[22] Some doubt is thrown on the estimated proportion of calories derived from food grains in China; if, for example, it were 90 per cent rather than 80 per cent, the Indian and Chinese estimates would tally.

These details aside, from what is generally known of the Chinese food situation, it is difficult to believe that conditions were as bad in 1965 as would be implied by a caloric intake of 1,800 per capita, compared with the none too high level of 2,110 in India in 1964. Even 2,000 calories seem rather low. The possibility of a more egalitarian distribution of food in China than in India may help to reconcile the difference, but the outright starvation and malnutrition represented by 1,800 calories appear not to have been prevalent since 1961. We conclude, then, that the Chinese peasants were enjoying an adequate diet in 1965, and from all accounts, in 1966 as well.

Very little is known of the supply of manufactured goods to the countryside in recent years. Until more information is forthcoming on this and other elements of the standard of living, the question of the relative welfare of Chinese workers and peasants must be left open. Both groups appear to be supplied with the basic necessities of life, but which one enjoys the larger sum of satisfactions is a rather subtle problem. One of the keys to it may be the services supplied by the state, to which we turn next.

State Welfare Benefits

A system of social insurance was initiated in the 1950's to supplement the cash income of industrial workers. The system covers employees of factories and mines employing 100 or more workers and employees, transportation

21. The greater importance of wheat in India and rice in China should not affect the comparison, since the caloric value per kilo of the two foods appears to be fairly close.

22. An adjustment should be made for the failure to include potatoes with grain in the Chinese food availability figure, due to lack of data. If the Taiwan pattern of consumption for 1965 were taken to represent all of China, about 160 calories should be subtracted on this account, lowering the total to be explained to 2,340.

and communications enterprises, and construction companies. Included are payments for industrial accidents, sickness and injury off the job, death benefits, old age pensions, and maternity benefits.[23] Administration is in the hands of the All-China Federation of Trade Unions and its constituent local bodies. Coverage rose from 20.3 per cent of the industrial labor force in 1952 to 46.9 per cent in 1957 (Schran, 1961, p. 311. The total number of persons insured was 13.7 million in 1958. Priestly, 1963, p. 89).

In 1956, state expenditures for health, pensions, and social welfare constituted 7.9 per cent of the total wage bill.[24] However, this figure cannot be taken to represent the supplementary income received by a covered worker, for the following reasons:

a. Social insurance covered only 30 per cent of the labor force in 1956.

b. Not all of the state social expenditures inured exclusively to the benefit of covered employees. Some of the health and welfare expenditures, such as public health and free medical care, went beyond the confines of the formal insurance schemes.

c. Welfare expenditures apart from social insurance were made by enterprises in the form of canteens, nurseries, and dormitories.

d. Rent, as we have already observed, absorbed only a minimal portion of the family budget due to the policy of subsidizing housing.

The supplement to wages from social insurance, medical care, and collective welfare, apart from rent, has been given as follows (Schran, 1961, p. 313):

1952—10.6 per cent   1955—15.1 per cent
1953—14.6   1956—16.5
1954—15.1

If we assume that consumption expenditures equal cash income, the rent subsidy in 1956 may be estimated at 6.5 per cent,[25] making the total supplement 23 per cent of wages.[26]

What did this mean in terms of services actually rendered? We have already examined the housing situation. On the health side, according to the best available estimate, there were between 50 and 75 thousand doctors trained in Western medicine in China in 1957, or 8,700 persons per doctor if the higher figure is accepted, and 1,800 persons per hospital bed (Orleans, 1961, p. 141). The comparable figures for India (1956) were 5,200 per doctor and 2,000 per bed (WHO, 1959). The Chinese were thus relatively worse off with

---

23. The latest available edition of the social insurance regulations is *Important Labour Laws and Regulations of the People's Republic of China*, 1961.

24. The total wage bill in 1956 was 13.6 billion *yuan*, while social expenditures were 1.064 billion *yuan*. N. R. Chen, 1967, pp. 446, 447, and 491.

25. Based upon the Communist estimate that 8.48 per cent of family expenditures among Shanghai workers went for rent in 1929–30 compared with 1.94 per cent in 1956. N. R. Chen, 1967, p. 436.

26. This is exclusive of state expenditures on education, which more than doubled from 1952 to 1956, and in the latter year, exceeded insurance and health payments by almost 180 per cent. N. R. Chen, *ibid.*, p. 446.

respect to medical personnel than the Indians, who were none too well off themselves. The 1957 levels in China represented a substantial improvement over 1952, when there were 11,000 persons per doctor and 3,200 persons per hospital bed (*Ibid.*).

The Chinese claimed in 1965 that 450,000 doctors and nurses had been trained during their first 15 years of power, 120,000 of them medical school graduates (JPRS, April 14, 1965). If we put the number of physicians in 1958 at 75,000, the claimed number of medical school graduates would imply a total of 138,000 physicians at the end of 1963,[27] reducing the number of persons per physician to about 5,000. Whether so many qualified physicians could have been trained during a period of economic crisis is at least open to question. However, enrollments in institutions of higher education rose very rapidly from 1958 to 1961 and, even though they may have dropped off later, the magnitude of the educational effort was great. The likelihood is that availability of physicians, as well as of other medical personnel, has improved since the middle 1950's, perhaps substantially.

It has been remarked that "one of the more stable elements of domestic policy has been the commitment of the Communist government to general health improvement. Health services are popular with the people, and this fact is a matter of some consideration even within an authoritarian system" (Joint Economic Committee, 1967, p. xiii). The improvement of health services, plus public health and hygiene campaigns, which have done a lot to clean up the cities, represent a considerable achievement by the Communist government in improving the quality of urban life. An Indian economist who visited China in the late 1950's made the following observations:

The third aspect of the country and the people that greet one forcibly is the extraordinary cleanliness and neatness everywhere compared to conditions before the Communists came to power. Public health and sanitation are attended to with positive vigor. There are practically no flies, no rats, no dogs, and no sparrows in China ... People no longer spit anywhere they like. The streets, the pavements, the kerbs, and the sidewalks are all swept clean and they stay clean, morning, noon, and evening. (Chandrasekhar, 1961, p. 24).

Hasty impressions can be misleading, but too many recent observers have come away from China with the same feeling about cleanliness and orderliness to leave much doubt of Communist accomplishments in this sphere.

The present coverage of social security schemes is not known. In 1958, some 14 million workers were eligible for benefits out of a nonagricultural labor force of 57 million persons (State Statistical Bureau, 1960, p. 218), but there has probably been considerable expansion since then. For one thing,

---

27. This is based upon the following assumptions: a) that the rate of attrition of physicians was 2 per cent a year, which Orleans felt was minimal; b) that the number of medical school graduates by the end of 1958 was 32,200, as reported by the Chinese, so that 87,800 were graduated from 1958 to 1964; c) that the number of graduates for each year between 1958 and 1964 was the same. Leo Orleans has estimated that there was 123,000 medical school graduates to the end of the 1963 school year. In Joint Economic Committee, 1967, p. 511.

though the old-age pension system first became effective in 1958, it would not have drawn an immediate flood of pensioners. To qualify, a man reaching the pensionable age of 60 would have to have been at work for 25 years, including five years in the enterprise from which he was retiring.[28] There is no system of unemployment insurance, because China, like other Communist countries, denies the existence of unemployment.

Industrial health and safety codes have been adopted, but there is no way of knowing whether they are enforced. The Communists claim to have brought about a great reduction in the rate of industrial accidents and disease compared with the old regime (*Labour Protection in New China*, 1960), but no statistics have been published to back up this claim. There is a basic eight hour day with shift work common in order to maximize the use of equipment. Labor turnover is reportedly low because of the scarcity of industrial jobs and the reluctance of workers to quit them (*Far Eastern Economic Review*, March 12, 1965, p. 439). The pattern of stability was undoubtedly disrupted by the Cultural Revolution, when workers were called out of the factories to aid one or the other of the contending factions, and it is not unlikely that those who guessed wrong in choosing sides found their jobs gone when they returned.[29]

The educational system expanded rapidly, at least up to 1960, with primary school enrollment reaching 90 million in that year. While urban elementary schools covered six years, many rural schools did not go beyond the third or fourth year. The following appraisal of the Chinese educational drive suggests that it is precisely in this area that the Communists have done most to promote social welfare:

It can be said that, except for periodic disruptions, Communist China has managed to create and operate an educational system that is ideally suited to her conditions and goals. Unable to provide the hundreds of millions of people with first-rate education, she has encouraged an atmosphere of learning, has made literacy among the masses one of the primary goals, has managed to elevate the overall educational level of rural youth, has trained adequate numbers of middle-level specialists and technicians, and, at the same time, has not neglected the economy's requirements for higher level professional personnel, particularly engineers and scientists. (Orleans, 1967, p. 518).

The closing of the schools in September 1966, and the difficulty of getting students back to the classroom in 1967, provides another example of the Chinese Communist propensity to disrupt successful institutions. The lowering of standards and the intimidation of teachers that marked the Cultural Revolution are bound to have adverse long run effects upon the educational system.

28. *Important Labour Laws and Regulations*, 1961, p. 21. For women, the pensionable age was set at 50, with 20 years of work experience and five years in the enterprise.
29. There were numerous accounts in the Chinese press in 1966 and 1967 of workers who had been given substantial sums of money out of enterprise funds in order to get them to participate in demonstrations and forays, and who subsequently repented and returned the money. The many object lessons cited suggest that the practice must have been widespread.

With all their limitations, these various types of collectively provided services appear to have been more important in providing a better life for the Chinese worker than increases in cash income. But since many of the services were provided by industrial enterprises rather than by the state directly, the distribution of the benefits must have been quite unequal. Those who were fortunate enough to work in the larger and more profitable enterprises with good medical dispensaries, canteens, kindergartens, housing, and other social amenities were far better off, regardless of cash income, than their colleagues in less profitable enterprises, or those in such trades as traditional transportation, who had no access to the welfare system at all. A recent book on China, which paints a generally rosy picture of social conditions there, makes the following observations about contemporary Nanking:

All day long, men, women and children pull loads along the main axial road through Nanking from the North Gate to the South Gate; and in the cool of the evening, their numbers increase. They pull carts with tree trunks, coal, oil drums, and cruelly huge rods of pig iron. There is a precise tally: a man alone pulls a load of three tree trunks, a man with a child helper five. One man, or one woman, pulls two iron rods, one grown-up and a child pulls three. They walk in halters, with a rag in one hand to wipe the sweat out of their eyes. There are also larger teams, for instance four men pulling a cart with two cable drums, and they look in control of things and of themselves, chanting in unison, and stopping at times for a cigarette with a lot of shouting and laughter. But the men and women alone, or with a child struggling halfway behind them in its little halter, are agonizing: an old woman, a bald man pulling iron with the veins standing out on his forehead, wiping his drawn face and only just at the beginning of the long avenue south, which slowly climbs upward.[30]

It should also be recalled that the social insurance system does not extend to the great majority of the Chinese people who live in the country. They are probably the recipients of better health and educational services than in pre-Communist China, but are worse off than industrial workers in access to collective services in general.[31] Since it is probably true that real income, apart from these services, is higher in the factories, China appears to be no exception to the general rule that industrial employees in underdeveloped countries constitute a minority elite, enjoying a higher living standard than the bulk of the population around them. The laws of economics seem inexorable, even for a country so little addicted to the market mechanism as China: great disparities in the level of productivity of factory and farm will

30. Hans Koningsberger, 1966, pp. 92–93. It should be pointed out these conditions prevailed long before the Communists came to power. The point is that the Communists have not eliminated them.
31. This was certainly true in the 1950's, as evidenced by the following statement from a speech by the Chairman of the Trade Union Congress: " ... the wages of apprentices, messengers and other manual workers, unskilled workers and those workers whose job is quite like that of peasants, were fixed a little too high, with the result that the peasants express disapproval about this. ... In regard to welfare facilities, we were a little too liberal in the provision of subsidies so that certain workers have got a wrong idea of relying solely on the state." *Eighth All-China Congress of the Trade Unions*, 1958, p. 44.

be reflected in income differences between workers and farmers, the words of Chairman Mao notwithstanding.[32]

Organization of the Labor Market

Chinese industrial workers, white collar and blue collar alike, are organized in the All-China Federation of Trade Unions. At the time of the Eighth Congress of the Federation in 1957, the Federation had 16.3 million members (1958, p. 27), with the total reportedly up to 20.8 million in 1963 (Priestly, 1963, p. 28). Total nonagricultural employment in 1957 was in the vicinity of 40 million (Emerson, 1965, p. 128). Of the latter total, however, there were about 11.5 million people in handicraft and other traditional enterprises, so that the degree of organization was almost 60 per cent. The organizational percentage was probably above the average in manufacturing, particularly in the larger enterprises, and below the average in the service sector.

The structure of the Chinese Federation was modeled upon that of the Soviet Union. There are 16 national unions, and the principle of organization is strict industrial unionism; there are no craft unions. At the provincial level, 27 regional federations coordinate trade union work. A local union committee must be set up in any enterprise or institution with ten or more members. There were reported to be 200,000 local units in 1954 (*March Toward Socialism*, 1956, p. 14), but the number must have expanded considerably since then.

Membership is not compulsory, but since social insurance and other trade union welfare programs are available only to members, the pressure to join upon those who are eligible for social insurance coverage is very great. The unions are financed by a check-off of 1 per cent of monthly wages, plus 2 per cent of the payroll turned over to them by the enterprise. In addition, enterprises are required to pay to the labor insurance fund 3 per cent of total payroll (*Important Labor Laws and Regulations*, 1961, pp. 10 and 13).

Both the early debate on the functions of trade unions in China, and the ultimate fate of the unions, are reminiscent of what happened in the first years of the Soviet Union. There were apparently those within the Chinese trade union bureaucracy, as in the early Soviet trade unions, who felt that, the new society notwithstanding, trade unions had an independent role to play in the protection of worker interests. Two basic questions were debated: "The first was whether the trade unions were required to accept the leadership of the Party; the second was whether the trade unions should regard production as their central task and how to handle correctly the relation between developing production and improving the workers' livelihood" (*Eighth All-China Congress*, 1958, p. 54).

---

32. During the "Hundred flowers" period in 1956 and 1957, when considerable peasant disatisfaction with their living standards were manifested, the government mounted a campaign to convince them that their living standards were not below those of workers. Hughes and Luard, 1959, p. 188. After the failure of the Great Leap and the emphasis on agriculture, the same theme was repeated.

It took ten years and the onset of Stalinism to resolve this question fully in the Soviet Union. The Chinese had less trouble. As early as 1953, trade union officials who were deemed guilty of an "economist tendency," of advocating wage equalitarianism, of spending labor insurance funds too lavishly, of placing excessive demands on management, of "engaging in a lot of empty talk about socialist [housing] standards instead of trying to meet the actual existing situation by adopting simpler and cheaper methods to provide more houses," were warned against "blind adventurism" and "overambitiousness." (*Seventh All-China Congress*, 1953, pp. 66–68). The General Program adopted by the 1957 Trade Union Congress called upon the unions to follow the leadership of the Communist Party and to help raise productivity by tightening labor discipline and spurring workers on to greater effort. Workers were to be taught "to recognize the unity of interests between the state and the individual and, when these two conflict, realize that individual interests should be subordinated to state interests" (*Eighth All-China Congress*, 1958, pp. 107–08). As for what might be termed the labor protectionist, as against the productivity, aspect of trade union work, the unions agreed that "on the basis of developing social production, the trade unions should *gradually* improve the material and cultural well-being of the workers" (*Ibid.*, p. 107. Italics ours). Their proper role was as a transmission belt carrying Party directives to the workers, (*Seventh All-China Congress*, 1953, p. 55), the standard Soviet conception.

As in the Soviet case, the trade unions were entrusted with administration of the social insurance system. In addition, they conducted a number of welfare programs at the enterprise level, including savings clubs, kindergartens, nurseries, libraries, and theaters. But their main job was to help improve working performance, largely through so-called socialist emulation campaigns (See Hoffmann, in C. M. Li, 1964). In the West, one of the principal functions of trade unions is to assist workers in prosecuting individual grievances against management. A procedure for settling such grievances was established in China in 1954, but how it works in practice is not known. Worker dissatisfaction with the representation they were getting from their unions came to the surface briefly during the "hundred flowers" period, but has since submerged.[33]

The Chinese trade unions have only a minimal role to play in the determination of wages. The Chinese industrial wage system was modeled upon that of the Soviet Union, with basic (usually eight) grade scales for each industry, and a ratio of about three to one between the highest and lowest grades. Workers are slotted into the wage structure on the basis of skill and

---

33. Hughes and Luard, 1959, pp. 115–118. Lai Jo-yu, the former chairman of the Federation, had the following to say about individual grievances: "We should support those demands which are reasonable, and help the management to settle them properly and as early as possible. In response to those demands which are unreasonable or, though reasonable, cannot be solved for the time being, we should teach the workers to consider things from an overall point of view and to have a correct understanding of the relationship between livelihood and production." *Eighth All-China Congress*, 1958, p. 46.

the nature of the job to be performed, with piece work and bonuses providing supplements to the basic rates.[34]

The interesting question to be asked about the wage system, apart from its implications for living standards, is whether and how well it has allocated the labor force among industries and enterprises. Dwight Perkins has advanced the view that:

> Wages have played only a very minor role in allocation of the urban labor force. The problem of allocating unskilled laborers among various enterprises does not exist because there have always been far more workers than jobs to fill, whereas with the highly skilled, the small number has made direct physical allocation a relatively simple matter. (Perkins, 1966, p. 146).

There is no doubt that the Communist regime practiced what amounted to compulsion in preventing migration from country to city, and, during certain periods, in directing surplus urban labor into village and farm work. The extent to which direction of labor was practiced within urban areas is not known. Even if it were true, however, that a small number of engineers and technicians were subject to job control, and that a good deal of unskilled labor was available, it does not necessarily follow that the allocative role of wages was "very minor." A case for the opposite conclusion can be made as follows:

a. In many underdeveloped countries there are large numbers of unskilled workers, but this does not mean that the supply schedule of qualified factory labor is infinitely elastic. To convert a coolie into a semi-skilled steel worker involves a considerable training cost, even assuming basic literacy to begin with. Not every enterprise could set up a training program, and it is not unreasonable to believe that workers were attracted from one enterprise to another by differences in earnings, including fringe benefits.[35]

b. Intra-enterprise skill differentials on the order of three to one not only provided an incentive for workers to improve their skills, in itself an allocative function, but it may well have been necessary to attract and hold properly qualified labor. Differentials of this order of magnitude are by no means uncommon in surplus labor economies operating under market wage conditions.[36]

---

34. Piece workers constituted 42 per cent of the labor force in 1956, but the number fell off during the Great Leap Forward. There is no doubt of the renewed importance of this method of wage payment after 1960, but the precise percentage of piece work is not known. It is worth noting that the Chinese never practiced piece work to the same extent as the Russians, perhaps because such schemes require more sophisticated enterprise administrative machinery than they had available, particularly in the earlier years.

35. Complaints that "young workers were promoted a little too fast compared with veteran workers, so the veteran workers express disapproval about this" suggest that market forces were operating in favor of younger and better trained personnel. *Eighth All-China Congress.* 1958, p. 44.

36. In Africa, a skill differential ratio of 7 to 1 is the norm. A recent comment on that area is not inappropriate to our argument: "Ideological preferences are strongly in favor of a compressed salary structure everywhere in Africa. Supply and demand conditions, however, favor a large spread." Berg, 1966, p. 201.

c. The Chinese employed a pattern of inter-industrial wage differentials similar to that of the Soviet Union, with heavy industry taking precedence over light industry (See Berg, 1964, p. 245). This could have been blind imitation or, more likely, a recognition that favored wage treatment was necessary to assure priority industry an adequate supply of qualified labor.

d. There is some scattered evidence, based mainly on travelers' accounts, that substantial wage differentials continued to prevail after 1960, when the Chinese could no longer be accused of copying the Russian model.[37]

None of this proves that the Chinese labor market operates according to the canons of competitive market theory; but, then, neither does any known labor market. It is our view, however, that, in the absence of evidence on large scale and persistent direction of labor, earnings did influence the flow of labor in Chinese industry to an appreciable extent.[38] This is not to say that the wages paid were economic in the sense that they reflected marginal product;[39] if they had, the skill differentials would probably have been larger. This would have been "economism," one of the worst sins in the Chinese Communist lexicon.

The trade unions may have a consultative voice at the national level when the basic wage scales are determined, but there is no collective bargaining in the Western sense. Their role in the wage determination process has been described in the following terms:

The trade unions take part in all decisions on wages. Representatives of the All-China Federation of Trade Unions take part in discussions with Government representatives in the formulation of the national wage plan, and the Government always consults with the trade unions in the promulgation of wage decrees.... The actual wages to be received by the individual workers, within the framework of the national wages policy, and the wages plan for the industry, are only fixed after full discussions between the trade unions and the management of the enterprises.... The trade

---

37. Wages were said to vary from a minimum of 40–50 *yuan* to a maximum of 110–120 *yuan* at a steel mill. *Far Eastern Economic Review*, March 12, 1965, p. 439. The Shanghai Textile Mill No. 2 was reportedly paying a range of 48 to 110 *yuan*. *Ibid.*, June 25, 1964, p. 633. The range at the Shenyang Textile Machinery Factory was 2 to 1. *Ibid.*, December 23, 1965.

38. Included in earnings would be fringe benefits and privileges of an economic character such as easier access to housing. In many countries (including the Soviet Union in the past) material benefits overshadowed cash wages, particularly at times of consumer goods stringency.

39. Perkins makes another observation which is at least doubtful; "There is little doubt that in China marginal productivity of unskilled and semiskilled laborers was well below their average productivity and, while the average increased each year, it is doubtful that there was any rise in marginal productivity at all owing to the constant excess availability of unskilled workers. Thus the average wage in China was higher than necessary from the point of view of allocative efficiency." 1966, p. 150. But this fails to take into account factor proportions. Most modern equipment has built into it relatively fixed labor requirements, so that even the availability of unskilled labor at very low cost would not affect the factor proportions used. The large enterprises built in China with Soviet help were probably largely of this character, and the marginal productivity of the workers in them may well have been *higher* than the average wage level. On the other hand, earnings in handicrafts and traditional occupations (including transfer payments) may have been above marginal product simply so that people could live—but this is not an unusual situation.

union representatives not only take part in the discussions to decide the amount of wages to be paid to each grade, and the differentials between them, but also the various categories of workers who will occupy the different grades. They also make recommendations and take part in discussions for promoting workers to a higher grade. In the case of jobs which are paid under the piece-work system, the wages to be paid are also decided through discussions between the trade unions and management. (*March Toward Socialism,* 1956, p. 35).

What this means in practice has undoubtedly varied from one enterprise to another and over time. The wage functions of Soviet trade unions could be described in precisely the same terms, and in Russia the importance of the unions has varied from complete impotence to significant negotiating power at the local level.

The trade union movement was an important pawn in the power struggle which underlay the Cultural Revolution. The All-China Federation had been attacked by the Maoists and there were reports that it was taken over in 1966 by a pro-Mao group. Liu Ning-yi, the chairman of the Federation, was among those accused of heresy, but in 1967 he still appeared to be occupying his union post (*New York Times,* March 10, 1967). There were reports from many cities of local union involvement in the factional struggle, usually in opposition to the Red Guards and in support of the local Communist Party leaders.

Whatever the eventual outcome, it is unlikely that the trade unions will undergo any fundamental change in either structure or function. They are useful cogs in the mechanism of control by which the industrial labor force is harnessed to the goals of increased output and productivity. When the greater availability of consumer goods permits living standards to be raised they may take on more of a consumptionist hue, as have the trade unions in the Soviet Union. But for the present and forseeable future, "the most important task of the trade union organizations is to unite and lead all the workers to struggle conscientiously and enthusiastically for greater and greater output" (*Trade Unions in People's China,* 1956, p. 34).

The Rural Wage System

Trade unionism and centrally determined wage schedules apply to urban employment, and most particularly to the industrial sector. A few words on the organization of the farm labor market are in order, although the information available is even more scattered than for the cities.

Most Chinese peasants belong to production teams, the basic units of agricultural organization. The effectiveness of the team's work is a major determinant of the level of income earned. The unit of wage accounting is the work point, with total compensation depending upon the number of work points earned. This piece work system is widely used as an incentive mechanism. It involves rating every farm task; in one case, for example, cutting stalks of ripe grain and bringing them to threshing was worth three points per 100 catties; threshing carried 1.2 points per 100 catties; tending a bullock, including housing and feeding it, was worth ten points a day (*Far Eastern Economic Review*, Feb. 27, 1967, p. 457). Quality controls are part of the system, with penalties for unsatisfactory performance. Where conditions of operation make individual accounting difficult, collective piece work is used.

The critical factor is the value of a work point, and this cannot be determined until the harvest is in and profit and loss calculations are made. Prior to that time, the peasants receive rations of food and basic necessities as well as a small amount of cash on account. During the period of more than a decade in which this system has been in use, the organizational unit for which profit accounting is made has been variously the commune, the production brigade, and the production team, the latter averaging from 20 to 40 households. Post-Leap practice had favored the smaller unit as a means of helping make reward for effort most visible to the individual, but there are portents in the Cultural Revolution of return to the commune as a part of Maoist ideology.

Every system of collective agriculture is faced with the problem of distributing income among peasants in such a way as to maximize individual effort, as a substitute for the age-old spur of necessity. Soviet agriculture works on the basis of labor days earned, a cruder incentive method than the Chinese, but sufficient for the lesser degree of labor intensity prevailing there. If the Chinese rural pay system operated in practice as it is supposed to in theory, it would come nearer than most to the desideratum of wages being proportional to the marginal product of labor.

Numerous deficiencies in the application of the system have been noted, however. Managerial costs have often been pushed too high by the complexity of the system, so that "cadres were unable to participate frequently in 'productive work' and thus became a burden on the 'producers'. Since some cadres were paid entirely out of collective income, this led to dissention, in some cases resulting in the payment of such low wages to cadres that they had little incentive to work efficiently" (Walker, 1968 pp. 427–28). The allocation of work points to the many different jobs on a farm is difficult, and there is evidence that "this detailed and difficult process of job evaluation and assessment was beyond the capacity of many collective cadres . . . There were many disputes over norms and rewards; peasants competed for jobs they considered were liberally valued and became intensely preoccupied with the number of work points obtainable from different tasks" (*Ibid.* pp. 428–29).

The Chinese Communists devised an ideally correct wage payment mechanism for their large farm sector. They had excellent instruments with which to apply it, the Chinese peasants, who, as the experience of Taiwan shows, are remarkably good economic calculators. What seems to have gone wrong was the managerial ingredient. Margins above subsistence were often too small to provide any real incentive, because of procurement price policy as well as natural conditions. The high degree of literacy which was so important a part of the Japanese agricultural success, and carried over into Taiwan as well, was absent on the Chinese mainland. The ideological considerations which dictated huge accounting units and blurred the connection between effort and reward helped contribute to the inefficiency of the system. Finally, the Cultural Revolution raised "economism," that is, rational economic calculation, to the status of a major sin. This will make it even more difficult to establish an income distribution system which will induce the Chinese peasant to farm the land as efficiently as he did his small private holdings in centuries past.

CHAPTER EIGHT

Foreign Economic Relations

Foreign trade has constituted for China an important means of facilitating and accelerating modernization. Particularly during the early years of the Communist regime, China lacked the facilities to produce the wide variety of machinery and equipment urgently needed for industrialization. Some raw materials were either not available in China, or available only in insufficient quantities. Foreign trade helped provide these necessary capital goods and raw materials.

Modern technology has been introduced into China largely through foreign trade. The Chinese acknowledged the benefits received from the importation of complete plants from the Soviet Union in the 1950's, and from Western Europe in recent years. Without imported capital goods, materials, and technology, the growth of the Chinese economy would have been much slower.

Foreign trade also provided a means of compensating for serious shortfalls in domestic production. The continued importation of food grains since 1961 is a case in point. In the early 1960's, the Chinese relied on foreign grain to relieve domestic food shortages. More recently, China has continued to import such relatively cheap food grains as wheat and barley, and at the same time has continued to export rice, a more expensive grain. This policy presumably yields comparative advantages.

The conduct of foreign trade has been influenced by political and ideological factors as well as by purely economic circumstances. Many examples can be cited, the most notable one being the decline of Chinese trade with the Soviet Union and Eastern Europe because of the Sino-Soviet dispute. A second example is the attempt made by China to use trade activities to set up de facto diplomatic relations with Japan. Much of China's trade with less developed countries in Asia, Africa, and Latin America has been carried on to promote Chinese influence in these areas.

Control and Organization of Foreign Trade

Given the objectives of Chinese foreign trade policy, it is understandable that the Communist government almost immediately set out to establish complete control of foreign trade. The proportion of total foreign trade conducted by the state rose from 61 per cent in 1950 to 99 per cent in 1955 (Mah, 1968, p. 674). Since 1956, foreign trade has been a state monopoly.

The principal agency for supervision and control of the foreign trade sector is the Ministry of Foreign Trade, set up in August, 1952. It was split off from the former Ministry of Trade, with domestic trade functions being given to a Ministry of Internal Trade. A number of state companies were organized under the Ministry of Foreign Trade, each importing and exporting a specific range of commodities or dealing with another specific aspect of trade, such as shipping. Distribution of imports and purchase of exports are carried on by these companies on the basis of a state plan. Trade negotiations are usually conducted by these companies as the operating units of the Ministry.

In addition to the Ministry of Foreign Trade and its subordinate agencies, there is the China Committee for the Promotion of International Trade, a government agency allegedly performing the functions of a chamber of international commerce (*Ibid.*, p. 675). It was organized to promote trade with other countries, particularly those having no diplomatic relations with China.

Foreign trade has always been an integral part of national economic planning. The procedure follows the general planning process described above. Briefly, control figures of imports and exports are determined by the State Planning Commission for long-term planning, and by the State Economic Commission for annual plans, on the basis of future requirements of capital goods and other materials. The control figures are then sent to the Ministry of Foreign Trade, which draws up specific import and export plans. After approval by the State Council and the National People's Congress, the Ministry of Foreign Trade assumes responsibility for directing and supervising the state import and export companies in executing the plans.

During the first FYP, export targets were generally determined on the basis of import requirements. In a report on foreign trade in 1954, the Minister of Foreign Trade stated the basic principle of trade planning in these terms: "Export is for import, and import is for the country's socialist industrialization" (Yeh Chi-Chuang, quoted in Mah, *Ibid.*). Domestic demand for foreign capital goods and raw materials determined the import targets; exports were for the purpose of paying for the imports. But for a wide variety of commodities there was frequent conflict between domestic demand and export requirements. An export quota was established for such essential commodities as food grains and edible oil. For most other commodities, however, priority was given to export, guaranteeing fulfillment of the export contracts and of the export plan (Mah, 1968, pp. 679–80).

Since agricultural products constituted the bulk of exports, the slow growth of agricultural output during the First FYP sometimes resulted in a serious shortage of export commodities. The problem became increasingly acute toward the end of the First FYP; the export plans were not fulfilled in 1956 and 1957, leading to a shortage of foreign exchange (*Ibid.*). The difficulty was aggravated by the termination of Soviet loans in 1957. In consequence, a new principle of foreign trade planning was adopted. For most imports, targets were to be set on the basis of the export plans, which in turn were to be determined by domestic supply and foreign demand. For certain essential goods, exports were to be geared to import requirements so that the necessary foreign exchange would be available in any event. An official statement made in 1964 indicated that the same principle was still operative (Wilson, May 14, 1964).

Trends in Foreign Trade

Official Chinese statistics on foreign trade are scanty. The few figures which have been published are of an aggregate nature covering only the years before 1958. Moreover, the utility of the published figures is questionable because of the use of disequilibrium foreign exchange rates. Arbitrarily set, these rates generally reflect neither changes in the domestic prices nor changes in the foreign trade prices of export or import commodities. The official *yuan*-dollar and *yuan*-ruble rates are not consistent with the ruble-dollar rate, thus creating a distorted picture of foreign trade volume, expressed in *yuan*. Western experts believe that the *yuan* is overvalued at the official *yuan*-dollar rate and undervalued at the official *yuan*-ruble rate (Eckstein, 1966, pp. 92–93, and Chao and Mah, 1964, 192–204). Over-valuation of the *yuan* versus the dollar would tend to understate the *yuan* value of China's trade with the Free World, while undervaluation of the *yuan* vis-a-vis the ruble would tend to overstate the *yuan* value of Sino-Soviet trade.

Because of the absence of reliable data on foreign trade emanating from Chinese sources, Western scholars have attempted to reconstruct independent estimates primarily on the basis of the trade statistics published by China's trading partners. Table VIII–1 presents three such estimates, valued in dollars. There are discrepancies among them stemming from differences in sources, in coverage (such as inclusion or exclusion of Asian Communist countries) and in the handling of such problems as double counting, valuation and currency conversion ratios. In spite of the discrepancies, however, the year-to-year changes in trade volume in the series move in the same direction, and some general conclusions can be drawn from them.

The economic development of China was clearly reflected in its foreign

TABLE VIII-1
CHINESE FOREIGN TRADE, 1950–1965
(in millions of U.S. dollars)

Year	Total	ECKSTEIN[a] Exports	Imports	Total	MAH Exports	Imports	Total	PRICE Exports	Imports
1950				1,553	699	854	1,210	620	590
1951	1,761	871	890	1,968	932	1,035	1,895	780	1,115
1952	2,146	1,039	1,107	1,929	940	989	1,890	875	1,015
1953	2,379	1,119	1,260	2,286	1,099	1,188	2,295	1,040	1,255
1954	2,666	1,345	1,321	2,498	1,197	1,301	2,350	1,060	1,290
1955	3,077	1,612	1,465	2,732	1,425	1,307	3,035	1,375	1,660
1956	3,006	1,615	1,391	3,138	1,691	1,446	3,120	1,635	1,485
1957	3,776	1,911	1,865	3,057	1,651	1,407	3,025	1,595	1,430
1958	4,232	2,221	2,011	3,834	1,973	1,861	3,735	1,910	1,825
1959	3,922	2,010	1,912	4,290	2,253	2,036	5,265	2,205	2,060
1960	2,985	1,571	1,414	3,938	2,011	1,926	3,975	1,945	2,030
1961	2,736	1,597	1,139	2,894	1,523	1,370	3,015	1,525	1,495
1962	2,970	1,699[b]	1,271[b]	2,601	1,525	1,075	2,675	1,525	1,150
1963				2,696	1,557	1,139	2,755	1,560	1,200
1964				3,322	1,930	1,392	3,245	1,770	1,475
1965							3,695[b]	1,955[b]	1,740[b]

Notes:
[a] Adjusted to take account of re-exports.
[b] Preliminary estimates.

Sources:
Eckstein, 1966, pp. 94–95.
Mah, 1968, p. 692.
Price, 1967, p. 584.

trade sector. There was rapid expansion from 1950 through 1959 except for the years 1952 and 1957, in which volume fell slightly as a result of a decline in import volume. The "Five-Antis" campaign, which interrupted economic recovery, and the embargo on strategic materials may have accounted for the 1952 decline. Over-investment in 1956 and shortages of agricultural products were probably the most important causes of import curtailment in 1957.

In 1958, the year of the Great Leap Forward, the volume of foreign trade rose sharply, about 25 per cent over the 1957 level. This reflected a large increase in import demand for capital goods to support the Great Leap programs and a substantial rise of agricultural output, permitting exportation of more agricultural commodities. Trade volume reached a peak in 1959.

During the period 1960–1962, the foreign trade data show a declining trend, reflecting the economic crisis which followed the Great Leap and the deterioration of Sino-Soviet relations. By 1962, trade volume appears to have declined to 50–62 per cent of the 1959 peak level. Economic recovery in 1962 was followed by a revival of foreign trade in 1963, and the upward trend continued during 1964 and 1965. However, 1965 trade volume was still about 14 per cent below the previous peak.

Statistical returns from China's trading partners indicate that during 1966 there were significant advances in trade volume over the 1965 level (Economist Intelligence Unit, 1967, pp. 9–10). But in 1967, the export trade dropped, probably as a consequence of the Cultural Revolution.[1] Exports to Hong Kong during June, July, and August, 1967, were 35 per cent below the total for the same period in 1966 (Economist Intelligence Unit, 1967, p. 7). The Communist disturbance in Hong Kong during the period undoubtedly played a major role in causing this decline.

Although the tempo of domestic economic development was the principal determinant of the level of foreign trade, Soviet credits in the early 1950's and the repayment of these credits in subsequent years were also of importance. China had an import surplus from 1951 through 1955; a large part of Chinese foreign trade was with the Soviet Union during this period, and China was able to finance the import surplus with Soviet credits. Beginning in 1955, however, China generated an export surplus in every year up to 1965, with the possible exception of 1960. Moreover, exports either grew faster than imports during the years of increasing trade (i.e., 1955–59 and 1963–65), or decreased more slowly than imports during the years of declining trade (i.e., 1960–1962). The reversion from import surplus to export surplus reflects the fact that, beginning in 1955, China was amortizing its debt to the Soviet Union. The export surplus seems to have continued into 1966 (*Ibid.*). But, in 1967, the decline in exports and the continued increase in imports

---

1. Chinese exports were estimated to have fallen by 9 per cent in 1967, although imports increased. The downward trend in exports appears to have continued in 1968. *New York Times,* July 14, 1968, p. 6.

apparently resulted once more in an import surplus (*Ibid.*, and *New York Times*, Sept. 21, 1967).

The Commodity Composition of Foreign Trade

The impact of domestic economic development upon the pattern of Chinese foreign trade may be examined more closely through the commodity composition of trade. Tables VIII–2 and VIII–3 present Eckstein's estimates for the years 1955 to 1963. We have already pointed out that the main purpose of the import trade during the 1950's was to acquire the capital goods essential to the industrialization program. In these years, the bulk of imports consisted of machinery, equipment and industrial materials. The share of machinery and equipment was over 40 per cent of total imports in 1959, but declined sharply during the post-Leap years to less than 10 per cent in 1962–63. With economic recovery and some purchases of complete plants from the West, the machinery and equipment percentage rose to 17 per cent in 1965 (Price, 1967, p. 584).

The expansion of industrial materials imports also reflected the pace of industrial growth. But the importation of such materials did not decline as sharply as that of machinery and equipment, and some materials were imported in even greater volume in both absolute and relative terms. Imports of metals, metal ores, and concentrates was curtailed in 1961, and then leveled off. Petroleum, however, continued to come in during 1961 even when the economy hit bottom, probably for military reasons (Eckstein, 1966, p. 109).

The importation of chemical products rose during the years of industrial expansion and declined in the crisis years. But the share of chemical imports in 1963 became larger than in previous years, partly because of a sharp increase in the import of fertilizers. Textile fibers gained relative to total imports after 1959, due to a sharp drop in domestic cotton production in the years of agricultural crisis.

The most dramatic change in the commodity structure of foreign trade was the sharp rise of food imports in 1961, to nearly one-third of the total from less than 3 per cent in the preceding years. This reflected clearly the serious magnitude of the agricultural crisis. The relative weight of foodstuffs rose even higher in 1962 and 1963, and may have still been 30 per cent of total imports in 1965 (Price, 1967, p. 586).

Some 70 per cent of the exports of pre-Communist China were agricultural in nature, with oil seeds (particularly soybeans and soybean products) heading the list, followed by textile fibers, tung oil, eggs and egg products, tea, and tobacco. Other export commodities included nonferrous metals, pig iron, coal, and a wide variety of handicraft products. Under the Communists, from 1955 through 1963, farm produce and textiles combined fluctuated between 62 and 74 per cent of total exports; pig iron and non-ferrous metals averaged about 10 per cent of the total; and the remainder was made up of a wide miscellany of products, including manufactured consumer goods

TABLE VIII-2
COMMODITY COMPOSITION OF CHINESE IMPORTS, 1955–63
(per cent)

	1955	1956	1957	1958	1959	1960	1961	1962	1963
1. Machinery and equipment	22.8	29.7	31.6	27.6	40.5	37.1	13.2	5.2	7.3
2. Metals, metal ores and concentrates	8.8	8.9	8.0	23.7	15.2	18.1	7.7	7.4	6.9
3. Petroleum and petroleum products	4.2	7.4	8.5	6.6	7.2	7.6	11.9	9.7	6.7
4. Chemical products	8.6	8.4	9.3	9.1	7.8	5.6	5.6	6.8	10.8
a) Fertilizer and insecticides	3.7	5.1	5.2	4.9	2.7	2.1	2.9	3.3	5.2
5. Textile fibers	8.0	6.5	9.5	7.6	5.6	9.1	10.0	9.5	15.1
6. Foodstuffs	2.1	1.9	2.3	2.4	0.3	2.5	32.5	39.4	36.3
7. Other	45.5	37.3	30.9	22.9	23.3	20.0	19.1	21.9	16.9
8. Total	100.0	100.0	100.0	100.0	100.0	100.0	100.0	100.0	100.0

Source:
Eckstein, 1966, pp. 106–107.

TABLE VIII-3
COMMODITY COMPOSITION OF CHINESE EXPORTS, 1955–63
(per cent)

	1955	1956	1957	1958	1959	1960	1961	1962	1963
1. Soybeans, oilseeds and products	17.4	13.9	10.7	7.8	8.8	7.6	2.9	3.1	3.1
2. Cereals	7.8	9.2	4.4	9.0	9.8	8.0	2.4	3.3	4.8
3. Livestock products	16.9	13.3	11.3	13.8	8.9	8.1	5.4	5.8	9.4
4. Fruits and vegetables	5.2	4.6	6.1	5.6	4.7	4.3	3.4	4.0	4.9
5. Tea and tobacco	5.4	4.9	5.8	4.8	4.1	3.1	2.3	1.8	1.8
6. Textiles									
a) Raw materials	8.0	7.2	6.4	3.9	7.1	6.3	4.7	3.4	4.3
b) Fabrics	7.5	10.6	13.2	12.2	15.0	18.9	22.8	21.8	20.4
c) Clothing and footwear	1.6	2.7	4.9	9.4	14.2	15.4	17.3	18.0	16.0
7. Industrial materials	18.7	21.4	22.2	18.8	14.0	15.2	15.7	15.1	12.4
8. Machinery and equipment	1.0	0.8	0.7	0.7	1.2	0.5	0.4	1.2	1.5
9. Other	10.4	11.3	14.4	14.0	12.2	12.7	22.7	22.5	21.6
10. Total	100.0	100.0	100.0	100.0	100.0	100.0	100.0	100.0	100.0

Source:
Eckstein, 1966, pp. 114–115.

other than textiles. This commodity structure was typical of an underdeveloped economy.

There was a significant shift during the Communist period, however, reflecting the growing importance of textile exports. Textiles (not including raw materials) rose from 9.1 per cent of exports in 1955 to 29.2 per cent in 1959, and averaged about 38 per cent from 1960 to 1963. This increase was achieved at the expense of farm produce exports, which declined from 60.7 per cent of the total in 1955 to 43.4 per cent in 1959, the year in which the absolute volume of agricultural exports reached a peak. The subsequent agricultural crisis resulted in both an absolute and relative decline. Farm products fell to 21 per cent of exports in 1961 and 1962, and then recovered to over 28 per cent in 1963.

By the early 1960's, textile products had replaced farm products as the leading export of Communist China. This does not mean that there had been an expansion in textile production. On the contrary, the decline in cotton output during the agricultural crisis resulted in a curtailment of the supply of raw materials, forcing textile plants to operate below capacity. The increase in textile exports after 1960 must have been accompanied by a reduction in domestic textile consumption.

Direction of Foreign Trade

Prewar China's main trading partners were Great Britain, Japan, the United States, Hong Kong, France, and Germany. Trade with the Soviet Union and Eastern Europe was of negligible importance. The geographic distribution was radically altered after 1949. Table VIII–4 presents data on the percentage distribution of foreign trade with Communist and non-Communist countries, and with the countries within the Communist bloc, from 1950 to 1965.

The outstanding feature is trade with the Soviet Union, which rose to a place of predominant importance in the 1950's and then declined sharply. Sino-Soviet trade was only 3 per cent of Chinese foreign trade in 1930 (Mah, 1968, p. 687); it rose to over 50 per cent from 1952 through 1955, and remained at 40 per cent or more in the latter half of the 1950's. This was a result of the "lean-to-one-side" policy of the Chinese Communist Party, the availability of Soviet credits, and the Western embargo on Chinese trade. The relative decline in the importance of Soviet trade after 1955 was due to the termination of the Soviet loans and the relaxation of Western trade controls. But the dominant position of the Soviet Union continued until 1961, when the direction of China's trade once again began to change drastically, partly because of the deterioration of political relationships with the Soviet Union, and partly because of the need for food from the West. The Soviet share of Chinese trade turnover declined consistently from 42 per cent in 1960 to 11 per cent in 1965.

Trade with Eastern Europe started from less than 2 per cent of total turnover in 1950 and rose to between 11 and 18 per cent during the rest of

TABLE VIII-4
DIRECTION OF CHINESE FOREIGN TRADE, 1950 TO 1965
(per cent)

| Year | Non-Communist countries | Total[a] | Communist countries | | | Cuba |
			Soviet Union	Eastern Europe	Asian Communist countries	
1950	71.1	28.9	26.9	1.6	0.4	
1951	48.5	51.5	39.6	10.8	1.1	
1952	30.5	69.5	51.0	16.9	1.6	
1953	32.3	67.7	50.7	14.8	2.2	
1954	26.2	73.8	54.0	15.8	4.0	
1955	25.9	74.1	56.0	14.3	3.8	
1956	34.1	65.9	47.1	14.9	3.9	
1957	36.0	64.0	43.2	16.5	4.3	
1958	37.1	62.9	40.6	18.0	4.3	
1959	30.6	69.4	48.3	15.4	5.7	
1960	34.5	65.5	41.9	16.1	6.4	1.1
1961	44.3	55.7	30.5	10.7	8.5	6.0
1962	47.3	52.7	27.9	8.6	9.8	6.4
1963	54.8	45.2	21.8	8.2	9.5	5.7
1964	65.3	34.7	13.8	7.5	7.0	5.5
1965	69.6	30.4	11.3	8.1	[b]	5.8

Note:
[a] Including trade with Yugoslavia.
[b] Not available.
Source:
Computed from data given in Price, 1967, pp. 583–608.

the decade. China's chief trading partners in Eastern Europe were East Germany and Czechoslovakia, accounting for about two-thirds of the total, followed by Poland and Hungary with 30 per cent (Price, 1967, p. 593). Chinese trade with Eastern Europe declined sharply in 1960, falling to 8 per cent by 1965. Trade with East Germany, Czechoslovakia, and Hungary, all of which supported the Soviet Union in the Sino-Soviet dispute, dropped more sharply than with the other European Communist countries. Trade with Rumania, which remained relatively neutral in the internecine Communist dispute, and with Albania, China's only ally in Eastern Europe, increased after 1960 (*Ibid.*, p. 596).

The level of foreign trade with North Korea, North Vietnam, and Outer Mongolia, began from less than 2 per cent of total turnover in 1950–52 and increased to nearly 10 per cent in 1962, when for the first time it exceeded Chinese trade with the European Communist countries. But it declined somewhat in 1964; with Outer Mongolia because of political disagreement between the two countries, and with North Korea because of the completion of drawings from Chinese credits granted in 1960 as well as the political leaning of North Korea toward the Soviet Union. On the other hand, Chinese trade with North Vietnam continued to increase; large-scale assistance was provided by China to support North Vietnam's First Five-Year Plan (1961–65) (*Ibid.*, p. 598).

Trade between China and Cuba began with a volume of $40 million in 1960, following the establishment of formal trade relations between the two countries, to a record volume of $213 million in 1965, and accounted for about 6 per cent of Chinese trade turnover in that year. Sino-Cuban trade involved mainly the exchange of Cuban sugar for Chinese food.

The rapid expansion of Sino-Soviet trade in the first half of the 1950's was accompanied by a decline in China's trade with non-Communist countries. The share of the non-Communist countries decreased from over 70 per cent in 1950 to 26 per cent in 1955. The relaxation of Western trade controls, the termination of Soviet loans, and the lack of a sufficient supply of industrial materials within the Communist bloc, led to some increase thereafter. But a sharp reorientation in Chinese foreign trade took place only after 1960, when the non-Communist portion rose to 70 per cent in 1965.

Chinese policy on trade with the West appears to be guided first and foremost by the need to acquire resources vital to economic development—machinery, equipment, industrial raw materials, and technical know-how. Securing essential food supplies has been a second consideration. Trade also is maintained with a number of countries primarily for the purpose of earning foreign exchange; Hong Kong, Malaya, and Singapore are important in this respect. Political considerations also motivates some trade.

China presently maintains direct or indirect trade relations with most countries outside the Communist bloc, but only a small number of them have played a significant role. Tables VIII–5 and VIII–6 show the relative importance of these countries in Chinese import and export trade from 1961 to 1964.

TABLE VIII–5

DISTRIBUTION OF CHINESE IMPORTS FROM NON-COMMUNIST COUNTRIES, 1961–64

(per cent)

	1961	1962	1963	1964
1. Australia, Canada and New Zealand	45.2	39.8	43.1	30.4
2. Japan	2.2	6.1	8.6	14.8
3. Argentina	0.6	5.0	0.5	10.4
4. Indonesia	4.1	6.1	4.8	6.3
5. United Kingdom	6.7	4.2	4.5	5.1
6. France	5.3	7.7	8.7	4.0
7. West Germany	5.9	5.5	2.3	1.9
8. Italy	4.9	3.5	2.7	1.9
9. Other	25.1	22.1	24.8	25.2
10. Total	100.0	100.0	100.0	100.0

Source:
Computed from data given in Price, 1967, p. 600.

TABLE VIII-6

DISTRIBUTION OF CHINESE EXPORTS TO NON-COMMUNIST COUNTRIES, 1961–64
(per cent)

	1961	1962	1963	1964
1. Hong Kong	20.5	22.8	23.0	24.3
2. Japan	5.2	7.3	9.6	14.4
3. Malaya and Singapore	9.6	10.6	12.2	9.1
4. United Kingdom	13.0	8.3	6.4	5.7
5. West Germany	6.3	5.3	4.6	4.7
6. Ceylon	3.8	3.3	4.3	3.8
7. Indonesia	7.1	7.6	4.6	3.7
8. Burma	3.8	4.5	3.4	3.3
9. France	2.3	2.5	2.6	2.7
10. Italy	1.8	2.0	2.6	2.0
11. Other	26.6	25.8	26.7	26.3
12. Total	100.0	100.0	100.0	100.0

Source:
Computed from data given in Price, 1967, p. 600.

On the import side, Australia, Canada, Argentina and Japan became China's most important trade partners; except for Japan, these countries were the major source of food import. Japan was the principal supplier of such industrial goods as steel products, fertilizers and other chemical products, machinery, synthetic fibers, motor vehicles, and earth-moving equipment. Other import partners were Indonesia (rubber), and the United Kingdom, France, West Germany, and Italy, all of which sent capital goods.

Hong Kong is the largest single purchaser of Chinese products by a substantial margin. China exports to Hong Kong not only many agricultural commodities, but also manufactured products, including textiles and machinery. Japan imported such bulk products as coal, pig iron, iron ore, salt, coke, and soybeans and foodstuffs. Malaya and Singapore have large Chinese populations and traditionally have purchased Chinese products. Other important Asian customers were Ceylon, Indonesia, and Burma. Four countries of Western Europe (the United Kingdom, West Germany, France, and Italy) purchased agricultural products and industrial raw materials.

A few general figures may help put the Chinese foreign trade sector in better perspective. Total trade volume (imports plus exports) was reported at 10.45 billion *yuan* in 1957 at current prices (N. R. Chen, 1967, p. 405). This can be broken down into imports and exports on the basis of the data in Table VIII-1; the resultant figures are 4.81 billion *yuan* of imports and 5.64 billion *yuan* of exports. Gross domestic product for 1957 at current prices has been estimated at 105.06 billion *yuan* (Liu and Yeh, 1965, p. 67). Imports thus constituted 4.6 per cent of the GDP, and exports 5.3 per cent, by no means negligible quantities.

The significance of foreign trade to individual sectors was still greater.

The value of industrial and mining imports in 1957 came to 2.25 billion *yuan*,[2] which was 10 per cent of the domestic product originating in manufacturing and mining (Liu and Yeh, 1965, p. 67). On the export side, agricultural exports constituted 6.4 per cent of the domestic product originating in agriculture (*Ibid.*).

Trade turnover rose by 22 per cent from 1957 to 1965 (Table VIII-1), and GDP by the same percentage (if the higher of the estimates in Table IX-1 is accepted). Both figures are extremely rough estimates, but they do suggest that the importance of foreign trade to the Chinese economy was much the same in 1965 as in 1957.

The Balance of International Payments

Tables VIII-7 and VIII-8 present estimates of the balance of payments of China for the period 1950 to 1964. By far the most important single item of international outlays in that period was for merchandise imports, which accounted for 80 per cent of total payments. Repayments of Soviet credits, foreign aid, and freight and insurance were next in relative importance, but together constituted only 19 per cent of total payments. Merchandise exports were the dominant source of international receipts, accounting for about 86

TABLE VIII-7
THE BALANCE OF INTERNATIONAL PAYMENTS OF CHINA
FOR THE PERIOD 1950 TO 1964

	Amount *(U.S. $ million)*	*Percentage* *distribution*
I. Payments, total	25,480	100.0
1. Merchandise imports, f.o.b.	20,326	79.7
2. Freight and insurance	1,135	4.5
3. Salary of Soviet experts	100	0.4
4. Expenditures of students in U.S.S.R.	70	0.3
5. Chinese expenditures abroad	150	0.6
6. Repayments of Soviet debts	2,244	8.8
7. Foreign aid	1,455	5.7
II. Receipts, total	25,981	100.0
1. Merchandise exports, f.o.b.	22,406	86.3
2. Foreign expenditures in China	90	0.4
3. Overseas remittances	1,050	4.0
4. Receipts from food parcels	141	0.5
5. Soviet credits	2,294	8.8
III. Balance	501	—

Source:
Mah, 1968, p. 723.

2. This figure was derived by applying the import-export ratios of Table VIII-1 and commodity data in Table VIII-2 to the total trade turnover of 10.45 billion *yuan* cited above.

TABLE VIII–8
YEARLY BALANCE OF INTERNATIONAL PAYMENTS OF CHINA,
1950 TO 1964
(millions of U.S. dollars)

Year	Payments	Receipts	Balance
1950	915	883	−32
1951	1,115	1,277	162
1952	1,058	1,574	516
1953	1,263	1,351	88
1954	1,568	1,647	79
1955	1,763	2,204	441
1956	1,913	1,820	−93
1957	1,917	1,740	−177
1958	2,336	2,048	−288
1959	2,520	2,330	−190
1960	2,352	2,097	−255
1961	1,732	1,685	−47
1962	1,429	1,642	213
1963	1,613	1,659	46
1964	1,985	2,023	38

Source:
Mah, 1968, pp. 722–23.

per cent of the total during 1950 to 1964, with Soviet credits contributing about 9 per cent more.

Changes in China's international financial position may be seen from Table VIII–8, which shows the yearly balance of payments for 1950 through 1964. A number of factors were responsible for the improvement that occurred from 1951 to 1955. The rapid growth of domestic production had increased the export capacity. Strict and effective centralized control of the economy, including trade and foreign exchange controls, enabled the government to channel domestic products to export and to allocate export earnings largely to the importation of capital goods. The payments position during this period was strengthened further by overseas remittances and long-term Soviet credit.

Deficits in the international accounts began to appear as early as 1956, largely because Soviet credits had dwindled at the same time as China was stepping up its foreign aid and debt repayment. The balance of payments deteriorated further in 1958 and 1959 when the Great Leap Forward stimulated the demand for imports. The subsequent sharp decline in agricultural exports forced China to take drastic measures in 1960, 1961, and 1962, to contain further deterioration of the balance of payments. Grain purchases in 1961 and 1962 were accompanied by a curtailment of the import of all other commodities.

Economic recovery in 1963 and 1964 made possible some expansion of both imports and exports. Although deficits still appeared in the international accounts, they were largely the result of debt repayments. By the end of 1964, China had liquidated nearly all of its remaining debt to the Soviet Union and East Europe. However, indebtedness to the rest of the world rose in 1963 and 1964 because of continued heavy drawings on credits for the purchase of grain and capital goods (CIA, 1967, pp. 621–60). China's holdings of Western currencies and gold at the end of 1964 totaled about 400 million dollars, compared with a reserve of 645 million dollars in 1957, when Chinese trade with Western countries was at a considerably lower level (*Ibid.*).

No solid estimates are available for China's balance of payments after 1965. Both exports and imports continued to expand in 1965 and 1966, with imports probably growing more rapidly than exports due to the completion of the repayment of the Soviet loans. The apparent result was a continued improvement in the nation's financial position. But China's payments position probably became unfavorable in 1967; incomplete trade data suggest that a sharp decline in Chinese exports may have taken place because of the disruption of domestic production by political turmoil.

The Chinese Foreign Aid Program

China embarked upon a foreign aid program in 1953. Given the fact that China itself had been deeply in debt, and that half the countries receiving Chinese aid had per capita national products roughly equal to or exceeding that of China, there was no logical economic reason for China to mount a foreign aid program. Yet the aid commitments to foreign countries during the period 1953 to 1965 totaled more than two billion dollars (Price, 1967, p. 589), close to the 2.2 billion dollars which the Soviet Union lent to China. Even during years of domestic economic difficulty, the foreign aid program seemed not to have declined significantly. During the economic crisis of 1961, aid commitments were extended in a larger amount than ever (Eckstein, 1966, pp. 306–07). The Chinese foreign aid program is primarily politically motivated, designed to promote Chinese influence. However, the possibility of long-run commercial gain cannot be ruled out.

The geographic distribution of China's foreign aid commitments during 1953–65 is shown in Table VIII–9. The bulk of the aid was extended to other Communist countries, particularly those in which Sino-Soviet competition for influence was most acute. North Vietnam and North Korea were the chief recipients. Outer Mongolia was another original recipient, but Chinese aid to that Soviet satellite was phased out in the first half of the 1960's (Price, 1967, p. 589). Albania began to receive Chinese aid in the mid-1950's and became a major recipient a decade later. Chinese aid was given to Cuba in 1960 and 1963 in the form of interest-free loans, amounting to a total of 100 million dollars (*Ibid.*). Chinese aid to Hungary was part of a bloc-wide program after the 1956 revolt.

Aid to non-Communist countries absorbed 40 per cent of total commitments, with half going to Asian neighbors. Indonesia was the chief recipient

TABLE VIII-9
CHINESE AID COMMITMENTS TO FOREIGN COUNTRIES, 1953–1965

	Amount (in millions of U.S. dollars)	Percentage distribution
I. Communist countries	1,223.5	60.0
1. North Vietnam	457.0	22.5
2. North Korea	330.0	16.2
3. Outer Mongolia	115.0	5.6
4. Albania	164.0	8.0
5. Cuba	100.0	4.9
6. Hungary	57.5	2.8
II. Non-Communist countries	815.0	40.0
1. Asia	410.0	20.1
2. Africa	264.0	13.0
3. Middle East	141.0	6.9
III. Total	2,038.5	100.0

Source:
Price, 1967, p. 589.

before its relations with China deteriorated. Other non-Communist Asian countries receiving Chinese aid included Burma, Pakistan, Cambodia, Nepal, and Ceylon. African nations received 13 per cent of total Chinese aid, or nearly one-third of the aid given to non-Communist countries. Algeria, Tanzania, and Ghana were the chief beneficiaries. The remainder of the Chinese aid went to the Middle Eastern nations of the United Arab Republic, Yemen, and Syria.

Limited by its economic and trade capabilities, the amount of aid that China could afford was very small compared with that given by the United States or the Soviet Union. As of 1964, Chinese aid to non-Communist countries constituted only 10–15 per cent of total Communist aid commitments (Eckstein, 1966, p. 237). But Chinese aid was concentrated in fewer countries than was the case with Soviet aid.[3] Moreover, China frequently showed more generosity in the terms of aid than did other donors. A significant proportion was in the form of free grants. Chinese credits were either interest free or given at very low interest rates, and provided for longer grace periods and periods of repayment. The Chinese argued that the underdeveloped nature of their economy enabled their aid program to provide technical assistance and commodity delivery much closer to the needs of the recipient countries than could be provided by advanced countries. Chinese aid programs were small quantitatively, but their qualitative aspects may have appealed to some of the less developed countries.

The extent to which the Chinese aid program may have created problems

[3]. Chinese aid was concentrated in 19 non-Communist countries while the Soviet Union extended aid to about 30 non-Communist countries. Eckstein, 1966, p. 238.

for the domestic economy is difficult to ascertain. In terms of gross national product, the amount of the annual expenditures for foreign aid was insignificant, probably amounting to no more than one-tenth of one per cent of the GNP (Kovner, 1967, pp. 609-20). The impact of aid expenditures on the living standards of the Chinese people may have been imperceptible, particularly because the bulk of the aid shipments were goods not for direct consumption.[4] On the other hand, the shipment of investment goods and industrial raw materials and the dispatch of technical and managerial personnel to foreign countries, may have deprived the Chinese economy of scarce resources badly needed for its own development.[5] In the long-run, the cumulative effects of the resource cost of the aid program may have been substantial. Whether the cost was matched by the benefits received is difficult to assess, not only because the economic cost of an aid program cannot easily be quantified, but even more because the benefits must also be judged on the basis of political objectives.

---

4. According to Eckstein, if all the aid shipments were assumed to be consumer goods, "total household consumption in the late 1950's probably could not have been reduced by more than 0.6 per cent." Eckstein, 1966, p. 166.

5. To the extent that capital exports included items which were underutilized in China, e.g., textile plants, the short-run opportunity cost would be reduced.

CHAPTER NINE

Prospects for the Chinese Economy

Economic Growth Since 1952

Various indicators of Chinese economic development are assembled in Table IX–1. All the figures, with the exception of the 1965 national product estimates, have been discussed in previous chapters, and require no additional comment here. The highly speculative national product estimates are considered below.

The impact of the First FYP (1953–1957) stands out in bold relief. Industrial production doubled during the period of the plan, despite the fact that the 1952 level was already well above prewar production. The impact was particularly noticeable in heavy industry, which far outpaced the consumer industries. There was also considerable progress in transportation, particularly in the expansion of the highway network. Foreign trade grew apace, reflecting in part the Soviet assistance program.

The weak spot was agriculture. The increase in grain output was small; in per capita terms, there was actually a decline because of the rapidly rising population. The cotton crop rose, but not fast enough to keep up with the growing demand for cloth.

Agricultural stagnation during the First FYP was largely a consequence of insufficient investment. Cultivated area increased by only 3.6 per cent. Unit-area yields showed some improvement for grain and cotton, but declined in the case of soybeans and major oilseeds.

Emphasis on industry did nothing to help solve the chronic Chinese problem of unemployment and underemployment. From 1952 to 1957, the Chinese population increased by 70 million. Non-agricultural employment grew by less than three million, and the ratio of non-agricultural employment to the total civilian labor force declined. The bulk of the increased population remained in the agricultural sector, aggravating population pressures and reducing the already meager capacity of agriculture to finance industrialization programs.

At the end of the First FYP, the Communist leadership, recognizing that industrial growth could not be sustained with a stagnant agriculture and increasing unemployment, began the Great Leap Forward. This strategy continued to lay emphasis on heavy industry, but in addition aimed at full utilization of underemployed rural labor in small industrial and farm projects. The economy made impressive gains in 1958, partly due to a good harvest. But, because of errors in planning and in plan implementation, the Great Leap led to a serious setback. While industry probably continued to grow until 1960, a severe farm crisis ensued. Net domestic product may have declined by as much as 15 per cent from a peak in 1958 to a trough in 1961.[1]

The economy began to recover in late 1961, after the abandonment of the Great Leap and the adoption of a new policy giving primary emphasis to agriculture. Beginning in 1962, the national product grew continuously, at least until the onset of the Cultural Revolution. Net domestic product in 1965 may have been about 7 per cent above the 1958 peak.[2]

Industrial production, which dropped sharply in 1961 and declined even further in 1962, began to recover in 1963. 1965 industrial production was probably slightly higher than the 1958 level. The recovery in industry was achieved by gradual re-employment of the capacity installed during and before the Great Leap Forward, rather than by new investment. Producer goods output recovered more rapidly than that of consumer goods. There were significant shifts in the relative importance of the various branches of industry. The most spectacular development was the rapid growth of the petroleum and chemical fertilizer industries, for which output in 1965 was perhaps five times as great as in 1957. The slow recovery of consumer goods was due to the failure of agriculture to provide an adequate supply of raw materials.

Foreign trade turnover, which had developed in the 1950's mainly in the context of the Sino-Soviet alliance, reached a peak of $4.3 billion in 1959, and then fell to a low of $2.7 billion in 1962. A rapid recovery of the export trade and the import of grains from the West pushed trade volume in 1965 up to $3.7 billion. But the commodity composition, particularly of imports, changed drastically. In the post-1960 recovery, nearly half of imports were in support of agriculture (grain, sugar, chemical fertilizer) in contrast to nearly 90 per cent in support of industry in the previous decade.

In view of the shift of national priority from industry to agriculture after

1. According to the exploratory estimate of Ta-Chung Liu, net domestic product at 1952 prices declined from 108 billion *yuan* in 1958 to 104.4 billion *yuan* in 1959, 95.9 billion *yuan* in 1960, and 92.2 billion *yuan* in 1961. 1967b, p. 50.

2. Our estimate of net domestic product of 116 billion *yuan* for 1965 is based on T. C. Liu's approach, with the assumption that grain production in that year was 200 million tons rather than the 185 million tons he uses.

the Great Leap, a rapid recovery of agriculture might have been expected. But the neglect of agriculture during the first decade of the Communist regime, and the extensive damage to the sector during the Great Leap period, made agricultural growth so slow and difficult that Chinese farming may be no more advanced now than in the early 1930's. To be sure, there have been important achievements in recent years. Increased use of chemical fertilizers and improved irrigation systems contributed to the rise of land productivity. But cultivated area in 1965 is believed to have been below the 1957 area. The output of major crops, with the exception of food grains, had not returned to the levels of the pre-Leap years. In per capita terms, crop production including grain, was considerably lower in 1965 than in 1957.

The Cultural Revolution, which started in the summer of 1966, undoubtedly had adverse effects on the economy. Their extent is difficult to determine. The available evidence indicates that the damage, although extensive, has not been nearly so great as the dislocations caused by the Great Leap Forward. Moreover, the impact on agriculture seems to have been much less than in industry and transport.[3]

In sum, Chinese economic development since 1949 has both its positive and negative aspects. For the economy as a whole, the annual rate of growth averaged 3.8 per cent during 1952–65.[4] China has built an industrial base capable of supporting a nuclear capability and producing a variety of important industrial products. On the other hand, agriculture has hardly kept up with population growth, and the country seems to be no nearer solution of its agricultural problem than it was when the Communists came into power.

China and India Compared

It is not easy to find a yardstick against which the performance of the Chinese economy can be measured. The only real candidate for this role is India, which approaches China in population and shares with its Asian neighbor a high degree of economic backwardness. Partial comparisons have been made in foregoing pages. It now remains to summarize them and to examine the aggregates. The relevant data are presented in Table IX–2.

The reader scarcely needs warning that the Chinese figures represent no more than informed guesses. This is no less true of the population statistics than of agricultural and industrial output. There may be wide margins of error, but the data do represent the consensus of Western analysts of the Chinese economy.

---

3. Chou En-lai is reported to have said that the targets for industry and communications were not met in 1967, and that output of many commodities was below 1966. "In the whole of last year, the chaos created by the capitalist-roaders, manipulation by the bad people, and sabotage by the enemy could not but affect our industrial and communications sector. Armed fights occurred, factory and mining equipment was damaged, and stoppages of work or slowdowns were reported." *New York Times,* April 25, 1968, p. 7.

4. This figure is substantially lower if the 200 million ton grain output figure is not accepted for 1965. See below, p. 222.

TABLE IX-1

SELECTED ECONOMIC INDICATORS FOR CHINA, 1933–65

Indicator	1933	1949	1952	1957	1965
I. *Population and employment*					
1. Population (millions)	500	545	575	645	728
2. Employment					
a. Non-agricultural employment (millions)	—	34.6	44.9	47.7	—
b. Proportion of non-agricultural employment to total civilian labor force (per cent)	—	14.9	14.1	13.4	—
II. *Net domestic product*					
1. Total (billion 1952 yuan)	59.5	—	71.4	95.3	108.1–116.0[a]
2. Per capita (1952 yuan)	119	—	126	150	160
III. *Industrial production*					
1. Index of industrial production (1956 = 100)	37.1	27.2	56.1	109.4	147.6
2. Output of major products					
a. Electric power (million kwh)	2,810	4,310	7,260	19,340	40,000
b. Coal (thousand tons)	28,379	32,430	66,488	130,000	210,000
c. Crude oil (thousand tons)	91	121	436	1,458	8,000
d. Crude steel (thousand tons)	410[b]	158	1,349	5,350	11,000
e. Chemical fertilizers (thousand tons)	—	27[c]	181[c]	803	4,600
f. Cotton cloth (million meters)	3,450[b]	1,889	3,829	5,050	3,900
g. Sugar (thousand tons)	415[b]	199	451	864	1,500
IV. *Agriculture*					
1. Area					
a. Cultivated acreage (thousand hectares)	102,318	97,881	107,919	111,830	109,000
b. Multiple crop index (per cent)	—	—	134.0	140.6	143.1
2. Yields (kilograms per hectare)					
a. Grains		1,063.5	1,374.8	1,530.8	1,600
b. Soybeans		611.3	815.3	788.3	833
c. Oilseeds (peanuts, rape, sesame)		652	790	650	750
d. Cotton		162.0	234.0	284.3	292

3. Output (million tons)					
a. Grains	173	108	170	185	200
b. Soybeans	11.8	5.1	9.5	10.1	7.5
c. Oilseeds (peanuts, rape, sesame)	6.4	2.3	3.7	3.8	3.4
d. Cotton	1.0	0.4	1.3	1.6	1.4
V. *Transportation* (kilometers)					
1. Railways	20,000	21,989	24,512	29,862	36,000
2. Highways	98,000	80,768	126,675	254,624	—
3. Inland waterways	—	73,615	95,025	144,101	—
4. Civil air routes	—	11,387[d]	13,123	26,445	—
VI. *Foreign trade volume* (million U.S. dollars)	515	1,210[a]	1,890	3,025	3,695

Notes:

[a] The higher estimate was derived by using Ta-Chung Liu's approach and the data he employed except that grain output was assumed to be 200 million tons in 1965.

[b] 1936 production figures.

[c] Not including ammonium nitrate.

[d] 1950 figures.

Sources:

Population and employment—Hou, 1968, pp. 356–57.
    E. F. Jones, 1967b, p. 93.

Net Domestic Product—T. C. Liu, 1967b, p. 50.

Industrial production—Chao I-wen, 1956.
    N. R. Chen, 1967, pp. 186–89.
    Field, 1967, pp. 293–95.

Agriculture—N. R. Chen, *ibid.*, pp. 284–87 and 338–39.
    E. F. Jones, *ibid.*, p. 94.
    Liu and Yeh, 1965.

Transportation—Kang Chao, 1968b p. 65.
    N. R. Chen, *ibid.*, p. 372.
    Economic Construction Quarterly, 1933, p. 234.

Foreign Trade—Y. K. Cheng, 1956, pp. 258–59.
    Price, 1967, p. 584.

TABLE IX-2

SELECTED INDICATORS OF ECONOMIC GROWTH,
CHINA AND INDIA, 1951–52 TO 1964–65

	CHINA	INDIA	
	Index 1965 (1952 = 100)	Index 1965 (1951 = 100)	Index 1964 (1951 = 100)
Population	128	135	132
National product	151–163[d]	171[a]	159[b]
National product per capita	118–127	127	120
Output of food grains	106–118	174	158
Industrial production	282	251	238
Steel	816	421	—
Chemical fertilizers	2,541	1,185	—
Cement	315	326	—
Cotton cloth	92	n.a.	125
Paper	278	401	—
Electrical capacity	722[c]	398[a]	—
Electric output	758[c]	503[a]	—
Railway trackage	147	106[a]	—

Notes:
[a] 1950–51 to 1964–65.
[b] 1950–51 to 1963–64.
[c] 1952 to 1964.
[d] See text for discussion of these figures.

Sources:
China—This volume, *passim*.
India—*Statistical Abstract of the Indian Union*, 1966.

The time span chosen for the comparison is, for China, from the year before the inception of the First FYP, 1952, to 1965, the last year for which the relevant estimates are available; and for India, from the beginning of its First FYP, 1951, to 1965. This means, for some of the comparisons, one year longer for India than for China. In general, except for population, adding a year for China would not affect the results greatly, since economic activity in 1966 does not appear to have been substantially greater than in 1965. However, the Indian end-year national product was very favorably affected by the bumper crop of 1964–65, when output was at a level not maintained the following year. For this reason, some data for 1964 have been included for India, since the 1963-64 harvest represented a plateau on which India had remained for three crop years.

The rate of population growth appears to have been much the same in the two countries for the period, somewhat in excess of 2 per cent per annum. Superimposed upon an absolute population base already high in relation to natural resources, this increase meant for both countries the need for strenuous efforts to avoid the Malthusian trap.

In evaluating the national product data, the factor of foreign assistance must somehow be taken into account. While it is true that national product presumably reflects the output of only those production factors that are owned by the nation itself, the foreign sector can affect appreciably the efficiency with which domestic factors are employed.

The only substantial source of assistance to China was the Soviet Union; its magnitude has been summarized by Eckstein as follows:

We know that the Soviets have made at least six long- or intermediate-term loans to Communist China. These include military loans of unknown value extended during the Korean War, two economic development loans of $300 million (1,200 million old rubles) and $130 million (520 million old rubles) granted in 1950 and 1954 respectively, two loans extended in 1954–55 for the purchase of the Russian share in the Sino-Soviet joint stock companies and of the Russian military stockpiles left in Port Arthur and Dairen, and finally a $45 million (180 million old rubles) sugar loan made in 1961. In addition ... in 1960 the Soviet Union either waived China's repayment obligation of $320 million (1,280 million old rubles) or granted a new loan for that amount. (Eckstein, 1966, pp. 129–160.)

This means that Soviet economic assistance to China amounted to between $475 million and $795 million, depending upon the interpretation of the 1960 loan. From 1950 to 1955, the loans (plus the military assistance) constituted between 2 and 6 per cent of government budget revenue; after 1955, they were not significant in this respect (Ecklund, 1966, p. 92). Moreover, by 1964, all but about $20 million had been repaid (Eckstein, 1966, p. 160).

India, in contrast to China, received substantial developmental aid throughout the entire period. The United States extended $2.8 billion in various forms of assistance from 1951 through 1964; plus $2.5 billion in wheat under Title I of Public Law 480, payable in blocked rupees (Chandrasekhar, 1965, p. 222). The Soviet Union provided India with about $1 billion of economic and military assistance, about half of which was still owing in 1965 (Goldman, 1967, pp. 105 and 111). Aid from all sources, including the United States and the Soviet Union, provided more than 2 per cent of domestic expenditures during the second Indian FYP, rising steadily thereafter to 4 per cent in 1964–65 (Mason, 1966, p. 21). Moreover, a large proportion of the debt incurred for developmental purposes by India was still outstanding in 1965, whereas China had virtually liquidated its indebtedness by then.

It would require a study of monographic proportions to determine quantitatively the advantage India obtained over China through the willingness of foreign countries to extend continued large-scale assistance to the former. The increase of gross investment by about one-quarter during its Second and Third FYP periods was attributable to foreign aid and certainly contributed to the Indian growth rate, but it is impossible to say by how much. The availability of P.L. 480 wheat may have created too much complacency with respect to agriculture and led to the postponement of serious efforts to increase agricultural productivity (Mason, 1966, p. 23). On the other hand, foreign aid provided India with critical foreign exchange and technical assistance, without which industrial development would have been greatly impeded. It is impossible, for these and other reasons, to assert that the impact of foreign aid on the rate of economic growth was proportional to its share in investment.

Some of the same considerations apply to China. Without Soviet aid and trade, Chinese industrialization would scarcely have been possible:

> From the gross economic point of view the Soviets were neither better nor worse off from the trade. Communist China, on the other hand, benefited enormously. The machinery imported from the U.S.S.R., if it could have been manufactured at all within China, could have been produced only at astronomical costs and with long delays. Such production would have required the scarcest and most costly manpower and industrial facilities within China. If there were to be any industrialization under these conditions, China would have to forego the consumption of great amounts of foodstuffs, textiles, and other domestically produced goods. But with the option of trade with the U.S.S.R. open, the Chinese could export a small quantity of foodstuffs, etc., to get the required industrial equipment. The cost of industrialization measured in these goods was an outstanding bargain in this case (Joint Economic Committee, 1967, p. 23).

Looking at the data in Table IX–2, our assessment of the comparative performance of China and India in terms of growth of the national product depends largely on interpretation of the agricultural record in both countries. For India, there is a considerable range between the good crop year of 1964–65 and the poor crop year of 1963–64. The argument for using the latter is that the favorable results of 1964–65 may have represented only a bulge between the stagnation of the three preceding years and the crop failure of 1965–66. On the other hand, the bumper crops of 1966–67 and 1967–68 suggest that 1964–65 may have been on an upward trend line which was temporarily interrupted by the unfavorable monsoons of 1965–66.

In the case of China, the lower estimate of GDP growth is that of Liu (Table VII–2), and is based critically upon a food crop estimate of 180 million tons in 1965. If one were to accept the figure of 200 million tons that some Western analysts feel is nearer the mark, the index of GDP growth in 1965 (1952 = 100) rises from 151 to 163.[5]

In terms of annual growth rates of the national product for the entire period under review, the following comparison emerges:

	China	India
High Chinese crop estimate/good Indian crop year	3.82%	3.90%
Low Chinese crop estimate/poor Indian crop year	3.25%	3.60%

5. The latter figures was derived by substituting 200 for 180 million tons of food grain in Professor Liu's model, everything else remaining unchanged except for the 1964 food grain figure, raised from 183 million tons to 195 million. Our grain estimates are derived from Table VII–4. See Ta-Chung Liu, "Quantitative Trends in the Economy," in Eckstein, Galenson, and Liu, pp. 87 *ff*.

The Chinese rate of growth is accounted for mainly by the period 1952 to 1958, when their national product rose by from 6 to 9 per cent a year (Joint Economic Committee, 1967, p. 51). The subsequent years contributed virtually no net growth. In India, on the other hand, GNP rose by about 3.5 per cent a year during the 1950's, and averaged 4.3 per cent a year during the first four years of the Third FYP (Mason, 1966, pp. 1–2). Indian growth was not as spectacular as the Chinese, but much steadier.

Examination of the sectoral data discloses equally sharp contrasts. China outpaced India in industrial growth, but did not do as well in agriculture.[6] The Chinese advantage in industry was offset by the higher growth rate of Indian agriculture.

The paucity of Chinese data limits the number of other direct comparisons that may be made. The available data suggest that expansion of the Chinese steel industry was double that of India, and that the Chinese did better in chemical fertilizers as well. On the other hand, Indian output of cotton cloth and paper grew more rapidly than that of China, reflecting once again the different development models chosen by the two countries. With respect to infrastructure, China clearly had the edge in the expansion of electric power, and also in growth of the railway network. However, India had a substantial advantage in both of these at the start of the planning period, and did not have to invest as much in building them up. India increased its network of surfaced roads by 67 per cent from 1951 to 1965, and showed a great expansion in both domestic and international aviation, but we do not know what the Chinese performance was in these areas, or in harbor installations, communications, and urban water and sewer facilities.

It is necessary once again to warn the reader that these comparisons are highly speculative. Even if the data actually reflected the situation more or less accurately, the time span is far too short for a firm judgment relating to such a long-term process as growth. Moreover, non-economic costs are not taken into account, and these may well be more important than the economic factors. Gross indexes of the type we have been obliged to use for lack of more detailed information can be misleading, and they must be employed with caution.

Future Policy Alternatives

In the final section of this concluding chapter, we suggest what in our view is an optimum general pattern for Chinese economic development. Before doing so, however, it may be useful to examine some of the implications of alternative policies.

6. Higher absolute acreage yields in China than in India should be considered. The rice yield per hectare in China (1952–57) was 2.54 tons compared with 1.36 tons for India (1955–1960). See Ishikawa, 1967a, p. 70. The question is whether it is easier to grow more rapidly from a lower or a higher absolute base. Intuitively, one would expect that it should be possible to advance more rapidly from the lower level, but this is not necessarily true until fairly high yields are attained.

One distinct possibility is a return to the big-push of the First FYP. This strategy, as will be recalled, was aimed at rapid industrialization, with the bulk of investment resources going to industry in general and to heavy industry in particular. During the First FYP, the ratio of gross capital formation to gross domestic product averaged 20–25 per cent; 48 per cent of investment was allocated to industry; and 85 per cent of industrial investment to heavy industry. Industrial growth was rapid, but agriculture suffered and eventually became a bottleneck that impeded further industrial growth.

Would it be possible for the Chinese to carry out successfully in the 1970's the same strategy they employed in the 1950's? To answer this question, we must look into the critical factor of the balance between food production and population.

The present agricultural base is hardly stronger than that of the 1950's. The current level of per capita grain output is in the neighborhood of 275 kilograms, compared with an average of about 290 kilograms during the years 1952 to 1957. If we accept a grain output estimate of 200 million tons and a population estimate of 728 million for 1965, and assume that during 1965–80, a) grain production will grow at alternative annual rates of from 1 to 4 per cent, and b) that population will grow at rates of from 1 to 2.5 per cent annually, the projected figures for per capita grain output in 1980, with varying combinations of grain and population increases, are as shown in Table IX–3.

TABLE IX–3
PROJECTED ESTIMATES OF PER CAPITA GRAIN OUTPUT IN 1980
(in kilograms per person)

Annual growth rate of grain output	ANNUAL GROWTH RATE OF POPULATION			
	1%	1.5%	2%	2.5%
1%	294	254	236	219
2%	318	295	274	255
3%	368	342	318	295
4%	426	396	368	342

The meaning of various levels of per capita grain availability has been described by Perkins as follows: "... it is difficult to avoid the conclusion that the Chinese population on the average in 1957 had about as much grain as they could directly consume. Further increases in output per capita would have pushed China into the ranks of countries which use large portions of their grain supply to feed cattle for meat or milk. This conclusion may seem odd given that grain was rationed in China in 1957. But a closer look at rationing in that year suggests that it was rather relaxed and people seem to have been able to buy all the grain they needed. Even a drop in per capita availability of some 20 per cent from the levels of 1957, for example, did not cause starvation in China, at least not on any scale" (1969, unpublished manuscript). Estimated grain availability in 1960 was 237 kilograms per capita.

Other analysts disagree with Perkins on the 1960 situation. Ashbrook's observations have already been quoted. Larson writes of "widespread famine" as a result of the crop failure (Joint Economic Committee, 1967, p. 221).

Whatever the precise situation, the 1960 level was about as low as a nation which relies mainly upon grain for nutrition could go without profound effects upon the health and industry of its people. A level of 290 kilograms represented a much more comfortable situation, but still one in which grain had to be consumed directly rather than converted into meat. A glance at Table IX-3 shows how closely attainment of the "comfort" level is linked with a favorable population development.

Despite priorities to agriculture, grain production rose by no more than 2 per cent per year during 1962–65. This rate would be difficult to maintain should China adopt a big-push strategy, which would mean shifting investment resources from agriculture to industry. If we make the optimistic assumption that under conditions of rapid industrialization grain output could continue to grow at an annual rate of 2 per cent throughout the 1970's, per capita grain output by 1980 would be between 255 kilograms and the present level of 274 kilograms, assuming that population continued to increase by between 2 and 2.5 per cent per annum, as in the past two decades. If population growth could be reduced, say to 1.5 per cent yearly, per capita grain output would have reached 295 kilograms by 1980. Abstracting from the possibility of large grain imports, which could only come through loans or a large volume of manufactured exports, a shift in investment priorities to industry would result in a food-population ratio in the coming decade distinctly worse than in the 1950's in the absence of a substantial decline in the rate of population growth. A forced industrialization program would thus be constrained, as it was in the First FYP, by the low level of agricultural production.

Moreover, some favorable factors that prevailed then are absent now. Soviet assistance, confiscation of Nationalist government and private capital, and coercive extraction of agricultural savings helped to finance the First FYP. These sources are no longer available. The population is now much larger than it was, implying more serious problems of unemployment and underemployment which cannot be resolved easily through industrialization.

Another basic alternative would be to continue the present policy of emphasis on agriculture, with a relatively slow rate of industrial growth. The goal might be to build in the course of two or three decades an agricultural base large enough to sustain a steady industrial advance. The success of this policy would depend on the ability of China to raise agricultural productivity, and to lower the rate of population growth. Referring once more to Table IX–3, we find that if a 4 per cent annual growth of grain output could be combined with 1.5 per cent population growth, there would be a substantial food surplus in 1980, by present standards.

To raise grain output by 4 per cent a year will be extremely difficult. Almost all the land worth cultivating has already been put under plow. Additional acreage can be obtained only at considerable expense. Greater output must be achieved primarily through increased productivity, which depends on more non-labor inputs. From 1962 to 1965, the use of chemical fertilizers rose by an estimated four million tons, while grain output increased by perhaps 20 million tons. If grain production were to increase by 4 per cent annually from 1965 to 1980, the crop in the latter year would reach 360 million tons, 160 million tons over the 1965 crop. This would entail an enormous expansion of current fertilizer production, even leaving aside the possibility of diminishing returns.

It will probably be easier to slow population growth than to raise farm productivity. But with properly formulated programs and sufficient additional resources, the hard-working Chinese peasantry might work wonders, as the experience of Taiwan has shown. The crux of the problem is the willingness of the Communist leaders to postpone industrialization.

A third policy might be greater reliance on foreign trade. Politics and ideology may prevent China from becoming a completely open economy. Expansion of foreign trade will be hampered by supply and demand elasticities, market imperfections, and state control of the trading organizations. Yet China has shown a flexible attitude toward foreign trade in the past. Trade volume has expanded to more than three times its 1949 level, and China now maintains trade relations with most of the world. In recent years, the Chinese Communists have not hestitated to utilize foreign trade to compensate for domestic shortages and to cultivate economic gains.

There are several lines along which China could conceivably pursue her comparative advantages. One would be to expand the import-substitution policy currently being followed, importing of low-yield wheat and exporting of high-yield rice.

A second approach would be to rely more heavily on foreign trade for the supply of technical agricultural inputs and to pay for them by greater farm exports. For example, the supply of chemical fertilizers is derived largely from domestic production. China imported complete modern plants from the West and established a large number of small plants in order to step up fertilizer production. But this is a costly and slow process, and direct imports of fertilizer would probably be more economical.

Another approach involves the development of labor-intensive manufacturing exports, which we discuss below. Considerable advantage could also be obtained by buying abroad certain advanced types of capital goods which China cannot at present produce in sufficient quantity to obtain economies of scale. High quality sheet steel provides a case in point (See

An Optimum Economic Policy for China

Let us suppose that the Civil war which has been raging under the name of the Cultural Revolution comes to a close, and that the Chinese leadership turns its attention to the task of moving the economy forward. What would be the elements of an ideal national policy designed to place the country on a path of steady and self-sustaining growth?

First and foremost, in our view, the rate of population growth must be reduced. Nothing would contribute more to easing the burden presently borne by the economy. The current rate of increase, on the order of 2 per cent per annum, means that there are 15 million more people to be fed each year. A heavy investment in birth control is called for.

That this is possible in the context of Chinese society is clear from the experience of Taiwan. Faced with an even more serious demographic problem, with its population swelling from 7.6 million in 1950 to 12.6 million in 1965, the Republic of China inaugurated a birth control program which reduced population growth from an annual rate of 3.8 per cent during 1951–1956 to 3.3 per cent from 1963 to 1965, and with indications of continued decline since 1965.

Next in importance is a comprehensive program to raise agricultural productivity. Changes in farm organization and intensification of labor are not the answer (See Ishikawa, 1967a, Chapter 3). Nor, as we have already indicated, can much be expected from bringing additional land under cultivation. "Crop production has been pushed about as far as traditional practices and methods will permit, and increased farm output can be attained only through the adoption of new technology, increased input, and improved practices" (Joint Economic Committee, 1967, p. 253). Such measures as greater use of chemical fertilizers, improved seed, more pesticides, and better farm implements are in order. The establishment of a comprehensive extension service is also essential if new techniques are to be exploited correctly.

Were China to open its doors and become an integral part of the world economy, crop specialization along lines dictated by comparative advantage would be an effective means of raising agricultural productivity. The experience of Taiwan is again instructive. The transformation there of an agricultural pattern based largely on sugar to one of large scale production of such cash crops as mushrooms, asparagus, pineapples, and bananas has been the principal factor behind the rapid economic growth of the island. With its huge manpower resources, China could become the world's supplier of labor-intensive agricultural commodities.

For much the same reason, the pattern of industrial development should be revamped. Given China's factor endowments, emphasis on heavy industry does not make economic sense at the present time. Labor-intensive industries such as clothing, textiles, electronics, and optical goods, which served Japan so well in an earlier stage of development and still absorb a great deal of employment, could be dominated by China. The expansion of the automobile, steel, shipbuilding, and heavy machinery industries should not be favored, since Chinese comparative advantage lies elsewhere. Small plants using a good deal of labor should be preferred to large factories.

The supply price of labor should stay low in China for many years to come, particularly since the necessary labor can be secured from the traditional urban sectors rather than requiring migration from farm to city. The correct pricing of capital offers a problem, but if the Chinese were to resort to world capital markets, as they have been doing to some extent, real capital costs would become clear.

The handicraft industries, which are labor-intensive by definition, have particularly good prospects for both the domestic and export markets. The Communists preserved the traditional crafts, but do not appear to have given sufficient attention to such factors as quality standards, design, and marketing organization, without which large scale commercialization cannot succeed. There is an enormous potential market for handicraft products. China could put art goods and domestic wares in millions of homes throughout the world, as well as in the country itself. But as the experience of India and many other nations has indicated, careful planning is required.

All of this hinges upon a great expansion of foreign trade. The volume of Chinese foreign trade turnover (imports plus exports) for 1965 has been estimated at $3.7 billion. India, with two-thirds the population of China, had a foreign trade turnover of $4.5 billion in the same year. The expansion would have to permit sharp changes in geographical direction and commodity composition. Since the break with the Soviet Union, the Chinese have been motivated largely by commercial advantage in their trade with the West. The great exception has been the United States, which appears now to be prepared to resume commercial relations with China.

Western Europe and the United States are the logical suppliers of the capital goods which China needs badly. These areas, in turn, must be prepared to absorb the products of labor-intensive industry which China is in a position to supply cheaply. This is the pattern of trade that is the nation's best hope for the next decade.

The events of the Cultural Revolution have left China in a state of economic drift, living from day to day with no apparent resolve to resume the drive for development. Sooner or later, a centralized authority will reassert itself and impose discipline on the economy. Any government, whatever its politics, will have to contend with the population problem and with the urgent necessity of raising farm yields.

Eventual policy toward industry is much more problematical. There have been some indications that the more extreme Maoists advocate a return to breakneck industrialization via the capital-intensive route. Autarky, rather than international trade, is consonant with traditional Chinese attitudes toward the outside world. A Great Power complex, with its military requirements, may induce the Communist leadership to go all out for heavy industry. Such a policy, whatever its contribution to modernization and growth, would mean for the Chinese people even greater hardships in the future than they have borne in the past.

Bibliography

The Agrarian Reform Law of the People's Republic of China and Other Relevant Documents, 1959. Peking: Foreign Languages Press.

Agricultural Economics Group, 1964. Economic Research Institute of Shantung Province, "A Survey Report on the Establishment of Stable- and High-Yield Fields by Ting-Chia Brigade in Hunang Hsien," *Economic Research,* No. 12, December 20, 1964, pp. 49–56.

Agricultural Economics Group, 1965. Economic Research Institute of Shantung Province, "Develop Water Conservation and Other Basic Construction for Farm Fields Through Labor Accumulation," *Economic Research,* No. 9, September 20, 1965, pp. 1–13.

Aird, John S., 1967. "Population Growth and Distribution in Mainland China," in Joint Economic Committee, *An Economic Profile of Mainland China,* Washington, D.C.: U.S. Government Printing Office, 1967.

Aird, John S., 1968. "Population Growth," in Eckstein et al., *Economic Trends in Communist China,* Chicago: Aldine Publishing Co., 1968, pp. 183–327.

"A Letter from Professor Alexander Eckstein to Chairman Proxmire, Dated April 20, 1967," in Joint Economic Committee, *An Economic Profile of Mainland China,* Washington, D.C.: U.S. Government Printing Office, 1967.

American Economic Review, Evanston, Illinois.

Ashbrook, Arthur G., 1967. "Main Lines of Chinese Economic Policy," in Joint Economic Committee, *An Economic Profile of Mainland China,* Washington, D.C.: U.S. Government Printing Office, 1967.

Ashton, John, 1967. "Development of Electric Energy Resources in Communist China," in Joint Economic Committee, *An Economic Profile of Mainland China,* Washington, D.C.: U.S. Government Printing Office, 1967.

Asian Survey, Berkeley, California.

Berg, Elliot, 1966. "Major Issues of Wage Policy in Africa," in A. Ross (editor), *Industrial Relations and Economic Development,* London, 1966.

Bergson, Abram, 1961. *The Real National Income of Soviet Russia Since 1928,* Cambridge: Harvard University Press, 1961.

Bibliography

Bergson, Abram, 1964. *The Economics of Soviet Planning*, New Haven: Yale University Press, 1964.

Bhalla, A. S., 1956. "Choosing Techniques: Handpounding v. Machine-Milling of Rice," *Oxford Economic Papers*, March, 1956.

Bhalla, A. S., 1964. "Investment Allocation and Technological Choice," *The Economic Journal*, Sept., 1964.

Buck, John Lossing, 1930. *Chinese Farm Economy*, Chicago: University of Chicago Press, 1930.

Buck, John Lossing, 1956. *Land Utilization in China*, New York: The Council on Economic and Cultural Affairs, Inc., 1956.

Buck, John Lossing, Owen L. Dawson and Yuan-li Wu, 1966. *Food and Agriculture in Communist China*, New York: Frederick A. Praeger, 1966.

Chandrasekhar, S., 1961. *Communist China Today*, Asia Publishing House, 1961.

Chandrasekhar, S., 1965. *American Aid and India's Economic Development*, New York, 1965.

Chang, John K., 1967. "Industrial Development of Mainland China, 1912–1949," *The Journal of Economic History*, 27, March, 1967, pp. 56–81.

Chang, Kuei-Sheng, 1961. "The Changing Railroad Pattern in Mainland China," *The Geographical Review*, 1961.

Chang, Kuei-sheng, 1963. *Petroleum Resources and Production in Mainland China*, Taipei: Institute of International Relations, 1963.

Chang, N. F., and H. L. Richardson, 1942. "Use of Soil Fertilizers in China," *Nature*, 49, No. 3780, April 11, 1942.

Chao Hsueh, 1957. "Problems Concerning Agricultural Mechanization in China," *Economic Research*, No. 4, April 9, 1957, pp. 16–18.

Chao I-wen, 1957. *New Industry in China*, Peking: Statistical Publishing House, 1957.

Chao, Kang, 1965. *The Rate and Pattern of Industrial Growth in Communist China*, Ann Arbor: University of Michigan Press, 1965.

Chao, Kang, 1967. *The Electric Power Industry in Communist China*, Washington, D.C.: Institute for Defense Analysis, June, 1967.

Chao, Kang, 1968a. "Policies and Performance in Industry," in Eckstein et al., *Economic Trends in Communist China*, Chicago: Aldine Publishing Co., 1968.

Chao, Kang, 1968b. *The Construction Industry in Communist China*. Chicago: Aldine Publishing Co., 1968.

Chao, Kang and Feng-hwa Mah, 1964. "A Study of the Rouble-Yuan Exchange Rate," *The China Quarterly*, No. 17, Jan.–March, 1964, pp. 192–204.

Chapman, Janet G., 1962. *Real Wages in Soviet Russia Since 1928*, Cambridge: Harvard University Press, 1962.

Chen, C. S., 1964. *The Mineral Resources and Industrial Capabilities of Manchuria*, unpublished manuscript, 1964.

Chen, H. S., 1932. *The Present Agrarian Problem in China*, Shanghai, 1932.

Chen, Nai-Ruenn, 1967. *Chinese Economic Statistics*, Chicago: Aldine Publishing Co., 1967.

Cheng, Chu-Yuan, 1957. *Income and Standard of Living in Mainland China*, Hong Kong, 1957.

Cheng, Chu-Yuan, 1963. *Communist China's Economy, 1949–1962*, Seton Hall University Press, 1963.

Cheng, Chu-Yuan, 1964. *Economic Relations Between Peking and Moscow*, New York: Frederick A. Praeger, 1964.

Cheng, Chu-Yuan, 1965. *Scientific and Engineering Manpower in Communist China*, Washington, D.C.: National Science Foundation, 1965.

Cheng, J. Chester, et al., 1966. *The Politics of the Chinese Red Army*, Stanford: The Hoover Institution on War, Revolution, and Peace, Stanford University, 1966.

Cheng, Yu-Kwei, 1956. *Foreign Trade and Industrial Development of China*, Washington, D.C.: The University Press, 1956.

Chi Ssu, 1960. "The Continued Leap Forward," *China Reconstructs,* 9, November, 1960, pp. 2–4.

Chi T'sun-wei, 1957. "How to Develop Chinese Industry More Proportionally," *Planned Economy,* No. 7, July, 1957, pp. 4–8.

Chi, Wen-shun, 1965. "Water Conservancy in Communist China," *The China Quarterly,* No. 23, July–Sept., 1965, pp. 37–54.

Chia An-nan, 1966. "Start from Reality, Produce and Promote Semi-Mechanized Farm Implements," *People's Daily,* January 4, 1966, p. 5.

Chiao Chi-ming, 1945. *Economics of Chinese Rural Society,* Chungking: Commercial Press, 1945.

Chin Ta-kai, 1965. "Problems of Communist Chinese Agriculture," *Research on Mainland China Problems,* No. 1, September 15, 1965.

China News Analysis, Hong Kong.

The China Quarterly, London.

China Report, New Delhi.

Chiu, A. Kaiming, "Agriculture," in H. F. NacNair (ed.), *China,* Berkeley and Los Angeles University of California Press, 1951.

Chu Chen-hua, 1966. "Several Problems Concerning Basic Construction of Farm Fields Through Labor Accumulation in the People's Communes," *Economic Research,* No. 2, February 20, 1966, pp. 28–33.

CIA (U.S. Central Intelligence Agency), 1960. *Average Annual Money Earnings of Workers and Staff in Communist China,* Washington, D.C.: U.S. Government Printing Office, October, 1960.

CIA, 1967. "Communist China's Balance of Payments, 1960–65," in Joint Economic Committee, *An Economic Profile of Mainland China,* Washington, D.C.: U.S. Government Printing Office, 1967, 2, pp. 621–60.

Clark, Colin, 1960. *The Conditions of Economic Progress,* 3rd edition, London: Macmillan, 1960.

Communist China, 1955–59; Policy Documents with Analysis, Cambridge: Harvard University Press, 1962.

Current Scene, Hong Kong.

Data Office, *Statistical Work Bulletin (Tung-chi Kung-tso Tung-hsin),* 1957. "Several Problems of Socialist Industrialization in China," *New China Semi-Monthly,* No. 1, Jan., 1957, pp. 67–71.

Department of Industrial Statistics, State Statistical Bureau, 1958a. *Major Aspects of the Chinese Economy Through 1956,* Peking: Statistical Publishing House.

Department of Industrial Statistics, State Statistical Bureau, ed., 1958b. *The Past and the Present of the Iron and Steel, Electric Power, Coal, Machinery, Textile and Paper Industries in China,* Peking: Statistical Publishing House, 1958.

Directorate General of Budget, Accounts, and Statistics, Republic of China, 1942. *A Statistical Analysis of Chinese Land Tenure System,* Chungking: Cheng-chung Publishing Co., 1942.

Eckaus, R. S., 1955. "The Factor Proportions Problems in Underdeveloped Areas," *American Economic Review,* 45, No. 4, Sept. 1955, pp. 539–635.

Ecklund, George N., 1966. *Financing the Chinese Government Budget,* Chicago: Aldine Publishing Co., 1966.

Eckstein, Alexander, 1964. "Sino-Soviet Economic Relations," in C. D. Cowan, *The Economic Development of China and Japan,* New York: Frederick A. Praeger, 1964.

Eckstein, Alexander, 1966. *Communist China's Economic Growth and Foreign Trade,* New York: McGraw-Hill Book Co., 1966.

Eckstein, Alexander, Walter Galenson and Ta-Chung Liu (ed.), 1968. *Economic Trends in Communist China,* Chicago: Aldine Publishing Co., 1968.

Economic Bulletin (Ching-chi Tao-pao), Hong Kong.

Economic Development and Cultural Change, Chicago.

Economic Journal, London.
Economic Research (Ching-chi Yeng-Chui), Peking.
Economist Intelligence Unit, 1967. *Annual Report to Quarterly Economic Review of China, North Korea, and Hong Kong,* London, 1967, No. 3.
Editor, *Current Scene,* 1966. "And Now There Are Four," 4, No. 2, Nov. 10, 1966.
Editor, *Economic Research,* 1956. "Why Have the Demand and Production of Double-Wheel and Double-Share Ploughs Stopped?" No. 9, 1956, pp. 1–4.
Eighth All-China Congress of the Trade Unions, 1958. Peking: Foreign Languages Press, 1958.
Elliot, Denis, 1965. "China's Waterways," *Far Eastern Economic Review,* 47, No. 2, January 14, 1965, pp. 50–51.
Emerson, John P., 1965. *Nonagricultural Employment in Mainland China, 1949–58,* U.S. Bureau of the Census, International Population Statistics Report, Series P-90, No. 21, Washington, D.C.: U.S. Government Printing Office, 1965.
Emery, Robert F., 1966. "Recent Economic Developments in Communist China," *Asian Survey,* 4, No. 6, June, 1966.
FAO (United Nations, Food and Agricultural Organization), 1966a. *Fertilizers: An Annual Review of World Production, Consumption and Trade,* 1965, Rome, 1966.
FAO, *Production Yearbook,* Rome, 1966.
Far Eastern Economic Review, Hong Kong.
Far Eastern Economic Review, Yearbook, 1962–66, Hong Kong.
Far Eastern Trade and Development, London.
Fei, Hsiao-tung, 1945. *Earthbound China,* Chicago: University of Chicago Press, 1945.
Field, Robert M., 1968. "Labor Productivity in Industry," in Eckstein et al., *Economic Trends in Communist China,* Chicago: Aldine Publishing Co., 1968, pp. 637–70.
Galenson, Walter, 1963. "Economic Development and the Sectoral Expansion of Employment," *International Labour Review,* 87, No. 6, June, 1963, pp. 1–15.
Gamble, Sidney D., 1933. *How Chinese Families Live in Peiping,* New York, 1933.
Goldman, Marshall I., 1967. *Soviet Foreign Aid,* New York, 1967.
Government of India, 1953. *First Five Year Plan,* New Delhi, 1953.
Government of India, 1956. Ministry of Food and Agriculture, *Report of the Indian Delegation to China on Agricultural Planning and Techniques,* New Dehli, 1956.
Government of India, 1961. Planning Commission, *Third Five Year Plan,* New Delhi, 1961.
"Great Victories in China's Economic Construction in 1960," 1961. *China Pictorial,* No. 2, Feb., 1961, pp. 2–3.
Han Chi-tung, 1945. *Estimated Chinese Losses and Damages during Sino-Japanese Hostilities, 1937–1943,* Nanking: Institute of Social Sciences, Academia Sinica, 1945.
Handicraft Section, Institute of Economic Research, Chinese National Academy of Sciences (ed.), 1957. *Nation-wide Survey Data on Individual Handicrafts in 1954,* Peking, 1957.
Harbison, Frederick and Charles A. Myers, 1964. *Education, Manpower, and Economic Growth,* New York, 1964.
Hirschman, Albert O., 1958. *The Strategy of Economic Development,* New Haven: Yale University Press, 1958.
Hitotsubashi Journal of Economics, Tokyo.
Ho, Franklin, L., 1951. "The Land Problem of China," *Annals of the American Academy of Political and Social Science,* No. 276, July, 1951, pp. 6–11.
Ho, Ping-ti, 1959. *Studies on the Population of China, 1368–1953,* Cambridge: Harvard University Press, 1959.
Hodgman, Donald, 1953. "Industrial Production," in Abram Bergson, ed., *Soviet Economic Growth,* Evanston: Row, Peterson and Co., 1953.
Hou, Chi-ming, 1965. *Foreign Investment and Economic Development, 1840–1937,* Cambridge: Harvard University Press, 1965.

Hou, Chi-ming, 1968. "Manpower, Employment, and Unemployment," in Eckstein et al., *Economic Trends in Communist China,* Chicago: Aldine Publishing Co., 1968, pp. 329–96.
Howe, Christopher, "The Supply and Administration of Urban Housing in Mainland China," *The China Quarterly,* January–March, 1968.
Hsia, Ronald, 1964. "Changes in the Location of China's Steel Industry," in C. M. Li, *Industrial Development in Communist China,* New York: Frederick A. Praeger, 1964.
Hsiao I-Shan, 1923. *A General History of the Ch'ing Dynasty,* Shanghai: Commercial Press, 1923.
Hsiao Yu, 1957. "How to Allocate Agricultural Investment," *Economic Research,* No. 9, Sept. 9, 1957, pp. 5–8.
Hsiao Yu, 1958. "Reclaim Waste Land and Expand Cultivated Acreage," *Planned Economy,* No. 2, Feb. 9, 1958, pp. 21–24.
Hsü Jung-an, 1965. "An Inquiry into the Problems of Building Stable- and High-Yield Farm Fields in China," *Economic Research,* No. 2, Feb. 15, 1965, pp. 15–19.
Hsü Ti-hsin, 1959. *An Analysis of the National Economy of China during the Transition Period,* revised edition, Peking, 1959.
Hseüh Mu-ch'iao, et al., 1960. *The Socialist Transformation of the National Economy in China,* Peking: Foreign Languages Press, 1960.
Huenemann, Ralph W., 1966. "Urban Rationing in Communist China," *The China Quarterly,* April–June, 1966.
Hughes, T. J. and D. E. T. Luard, 1959. *The Economic Development of Communist China,* London: Oxford University Press, 1959.
ILO (International Labor Office), 1964a. *Employment and Economic Growth,* Geneva, 1964.
ILO, *Yearbook of Labor Statistics,* 1957, 1964b, and 1967, Geneva.
Impartial Daily (Ta-kung Pao). Peking and Hong Kong.
Important Labour Laws and Regulations of the People's Republic of China, Peking: Foreign Languages Press, 1961.
International Journal of Agrarian Affairs, London.
International Labour Review, Geneva.
Ishikawa, Shigeru, 1965. "Strategy of Foreign Trade Under Planned Economic Development," *Hitotsubashi Journal of Economics,* January, 1965.
Ishikawa, Shigeru, 1966. "Choice of Techniques and Choice of Industries," *Hitotsubashi Journal of Economics,* February, 1966.
Ishikawa, Shigeru, 1967a. *Economic Development in Asian Perspective,* Tokyo, 1967.
Ishikawa, Shigeru, 1967b. *Factors Affecting China's Agriculture in the Coming Decade,* Tokyo: The Institute of Asian Economic Affairs, 1967.
Johnston, Bruce F. and John W. Mellor, 1961. "The Role of Agriculture in Economic Development," *American Economic Review,* September, 1961.
Joint Economic Committee, U.S. Congress, 1965. *Current Economic Indicators for the U.S.S.R.,* Washington, D.C.: U.S. Government Printing Office, 1965.
Joint Economic Committee, U.S. Congress, 1967. *An Economic Profile of Mainland China,* Washington, D.C.: U.S. Government Printing Office, 1967.
Jones, Edwin F., 1966. *China's Economic Revolution,* processed, 1966.
Jones, Edwin F., 1967a. "The Emerging Pattern of China's Economic Revolution," in Joint Economic Committee, *An Economic Profile of Mainland China,* Washington, D.C.: U.S. Government Printing Office, 1967.
Jones, Edwin F., 1967b. "Comment on Remarks by T. C. Liu and Kang Chao," in *Mainland China in the World Economy,* Washington, D.C., 1967.
Jones, P. H. M., 1964. "Creeping Modernization," *Far Eastern Economic Review,* 49, No. 7, November 12, 1964, pp. 350–52.
Jones, P. H. M., 1965a. "Advance in Agriculture," *Far Eastern Economic Review,* 49, No. 14, September 30, 1965, pp. 613–615.
Jones, P. H. M., 1965b. "The Farmer's Year," *Far Eastern Economic Review,* 50, No. 12, December 23, 1965, pp. 566–67.

The Journal of Asian Studies, Ann Arbor, Michigan.
JPRS (U.S. Joint Publications Research Service), 1965. *Communist China Digest,* No. 145, April 14, 1965.
JPRS, 1966. *Translations from Communist China: Economic,* No. 7, Aug. 10, 1966.
JPRS, 1967. *Communist China Digest,* No. 184, May 25, 1967.
JPRS, 1967. *Translations on Communist China,* No. 29, July 17, 1967; and No. 34, Nov. 7, 1967.
Kaplan, Norman, 1953. "Capital Formation and Allocation," in Abram Bergson (ed.), *Soviet Economic Growth,* Evanston: Row, Peterson and Co., 1953.
Kaplan, Norman, 1963. "Capital Stock," in Abram Bergson and Simon Kuznets, *Economic Trends in the Soviet Union,* Cambridge: Harvard University Press, 1963.
Kaye, William, 1961. "The State of Nutrition in Communist China," *The China Quarterly,* July-Sept. 1961.
Kershaw, Joseph A., 1953. "Agricultural Output and Employment," in Abram Bergson (ed.), *Soviet Economic Growth,* Evanston: Row, Peterson and Co., 1953.
Klatt, Werner, 1961. "Communist China's Agricultural Calamities," *The China Quarterly,* No. 6, April-June, 1961.
Klatt, Werner, 1967. "Comment on Dwight H. Perkins, Economic Growth in China and the Cultural Revolution," *The China Quarterly,* No. 31, July-September, 1967.
Koningsberger, Hans, *Love and Hate in China,* New York, 1966.
Kovner, Milton, 1967. "Communist China's Foreign Aid to Less-Developed Countries," in Joint Economic Committee, *An Economic Profile of Mainland China,* Washington, D.C.: U.S. Government Printing Office, 1967.
Kuo, Leslie T. C., 1964. "Agricultural Mechanization in Communist China," *The China Quarterly,* No. 17, Jan.-March, 1964.
Kuznets, Simon, 1957. "Quantitative Aspects of the Economic Growth of Nations," *Economic Development and Cultural Change,* 5, Supplement, No. 4, July, 1957.
Kuznets, Simon, 1961. "Economic Growth and the Contribution of Agriculture: Notes on Measurement," *International Journal of Agrarian Affairs,* 3, No. 2, April, 1961.
Kwang, Ching-wen, 1966. "The Economic Accounting System of State Enterprises in Mainland China," *The International Journal of Accounting,* 1, No. 2, Spring, 1966, pp. 61–99.
Labour Protection in New China, 1960. Peking: Foreign Languages Press, 1960.
Lamer, M., 1957. *The World Fertilizer Economy,* Palo Alto: Stanford University Press, 1957.
Landsberg, Hans H., 1964. *Natural Resources for U.S. Growth,* Baltimore: Johns Hopkins Press, 1964.
Larson, Marion R., 1967. "China's Agriculture under Communism," in Joint Economic Committee, *An Economic Profile of Mainland China,* Washington, D.C.: U.S. Government Printing Office, 1967.
Leibenstein, Harvey, 1957. *Economic Backwardness and Economic Growth,* New York: John Wiley & Sons, Inc., 1957.
Lewin, Pauline, 1964. *The Foreign Trade of Communist China,* New York: Frederick A. Praeger, 1964.
Lewis, John P., *Quiet Crisis in India,* New York: Doubleday & Co., Inc., 1964.
Li, Choh-Ming, 1951. "International Trade," in H. F. MacNair (ed.), *China,* Berkeley and Los Angeles: University of California Press, 1951, pp. 492–506.
Li, Choh-Ming, 1959. *Economic Development of Communist China,* Berkeley: University of California Press, 1959.
Li, Choh-Ming, 1962. *The Statistical System of Communist China,* Berkeley: University of California Press, 1962.
Li, Choh-Ming (ed.), 1964. *Industrial Development in Communist China,* New York: Frederick A. Praeger, 1964.
Li Fu-ch'un, 1957. "Report at the Sixth National Statistical Conference," *Statistical Work Bulletin,* No. 22, Nov., 1957.

Liao Lu-yen, 1955. "Report on the Basic Condition of Agricultural Production in 1954 and Measures to Raise Present Production," in *Rural Work Questions in 1955,* Peking, 1955, pp. 10–23.

Lieu, D. K., 1927. *China's Industries and Finance,* Peking, 1927.

Lieu, D. K., 1948. *China's Economic Stabilization and Reconstruction,* New Brunswick: Rutgers University Press, 1948.

Lieu, D. K., and Chung-min Chen, 1928. "Statistics of Farmland in China," *Chinese Economic Journal,* March, 1928, pp. 181–213.

Lin Hung, 1964. "To Establish Gradually Stable- and High-Yield Farm Fields Is the Major Way to Develop Agricultural Production in China," *Economic Research,* No. 9, Sept. 15, 1964, pp. 12–21, 26.

Lin Pin, 1966. "Agricultural Trends in Communist China," *Studies on Chinese Communism,* 9, No. 4, April 30, 1966, pp. 1–6.

Liu, Jung-Chao, 1965a. "Fertilizer Application in Communist China," *The China Quarterly,* Oct.–Dec., 1965, pp. 28–52.

Liu, Jung-Chao, 1965b. "Fertilizer Supply and Grain Production in Communist China," *Journal of Farm Economics,* 47, No. 4, Nov., 1965, pp. 915–932.

Liu Sheng, 1965. "On Several Basic Problems Regarding Water-Conservancy Construction in China," *Economic Research,* No. 6, June 20, 1965, pp. 21–26.

Liu, Ta-Chung, 1946. *China's National Income 1931–36: An Exploratory Study,* Washington, D.C.: The Brookings Institution, 1946.

Liu, Ta-Chung, 1967a. "Two Estimates of Grain Production, 1960–1965," in *Mainland China in the World Economy,* Washington, D.C., 1967.

Liu, Ta-Chung, 1967b. "The Tempo of Economic Development of the Chinese Mainland, 1949–65," in Joint Economic Committee, *An Economic Profile of Mainland China,* Washington, D.C.: U.S. Government Printing Office, 1967.

Liu, Ta-Chung and Kung-Chia Yeh, 1965. *The Economy of the Chinese Mainland: National Income and Economic Development, 1933–1959,* Princeton, N.J.: Princeton University Press, 1965.

Lo Wen and Shang-kuan Chang-chun, 1957. "The Problem of Expanding Irrigation in Farm Fields," *Economic Research,* No. 9, October 9, 1957, pp. 15–17.

Ma Yin-chu, 1958. *My Economic Theory, Philosophical Thought and Political Position,* Peking: Finance Publishing House, 1958.

MacDougall, Colina, 1965a. "Fertilizer Drive," *Far Eastern Economic Review,* 49, No. 1, July 1, 1965, pp. 14–16.

MacDougall, Colina, 1965b. "Keeping House in China," *Far Eastern Economic Review,* 50, December 23, 1965.

MacDougall, Colina, 1966a. "China's Foreign Trade," *Far Eastern Economic Review,* 51, No. 4, January 27, 1966, pp. 121–125.

MacDougall, Colina, 1966b. "The Pig and the Peasant," *Far Eastern Economic Review,* 52, No. 3, April 21, 1966, pp. 153–56.

Mah, Feng-hwa, 1961. "The Financing of Public Investment in Communist China," *The Journal of Asian Studies,* 21, No. 1, November, 1961, pp. 33–48.

Mah, Feng-hwa. 1968. "Foreign Trade," in Eckstein et al., *Economic Trends in Communist China,* Chicago: Aldine Publishing Co.

Malenbaum, Wilfred, 1959. "India and China: Contrasts in Development Performance," *American Economic Review,* June, 1959.

Malenbaum, Wilfred, 1962. *Prospects for Indian Development,* Glencoe, Ill.: Free Press, 1962.

March Toward Socialism, 1956. Peking: Foreign Languages Press, 1956.

Mason, Edward S., 1966. *Economic Development in India and Pakistan,* Cambridge: Harvard University Press, 1966.

Meier, Gerald M., 1964. *Leading Issues in Development Economics,* London: Oxford University Press, 1964.

Bibliography

Ministry of Agriculture, People's Republic of China, 1958. *A Collection of Statistical Data on Agricultural Production of China and Other Major Countries in the World*, Peking: Agricultural Publishing House, 1958.
Ministry of Economic Affairs, Republic of China, 1943. *Industrial Statistics of the Inland Provinces*, Chungking, 1943.
"Model Regulations for Agricultural Producer Cooperatives," 1955. *New China Monthly*, 74, No. 12, 1955, pp. 141–149.
Moyer, Raymond T. et al., 1947. "Farm Tenancy in China," *National Construction Journal*, 9, No. 1, July, 1947, pp. 13–24.
Munthe-Kaas, Harold, 1966. "China's Fields and Factories." *Far Eastern Economic Review*, 6, No. 5, Feb. 3, 1966, pp. 153–55.
National Program for Agricultural Development, 1956–1967 (Draft), Peking: People's Press, 1956.
New China Monthly (Hsin-hua Yueh-pao), Peking.
New China Semi-Monthly (Hsin-hua Pan-yueh-kan), Peking.
New York Times, New York.
Nove, Alec, 1961. *The Soviet Economy*, New York: Frederick A. Praeger, University Series, 1961.
Orleans, Leo A. *Professional Manpower and Education in Communist China*, Washington, D.C.: National Science Foundation, 1961.
Orleans, Leo A., 1967. "Communist China's Education: Policies, Problems and Prospects," in Joint Economic Committee, *An Economic Profile of Mainland China*, Washington, D.C.: U.S. Government Printing Office, 1967.
Ou, Pao-san, 1946. "A New Estimate of China's National Income," *Journal of Political Economy*, Vol. LIV, No. 6, December, 1946, pp. 547–554.
Ou Pao-san, et al. 1947a. *China's National Income, 1933*, 2 vols., Shanghai, 1947.
Ou Pao-san, 1947b. "China's National Income, 1933, A Revision," *Social Science Journal*, 9, No. 2, December, 1947, pp. 130–133.
Ou, Pao-san, and Foh-shen Wang, 1946. "Industrial Production and Employment in Pre-War China," *Economic Journal*, 56, No. 223, Sept., 1946, pp. 426–434.
Pauley, Edwin W., 1946. *Report on Japanese Assets in Manchuria to the President of the United States*, Washington, D.C.: U.S. Government Printing Office, July, 1946.
Peng Tse-yi (ed.), 1957. *Materials on the Modern History of Handicrafts in China*, Peking, 1957.
People's Handbook for 1964, Peking: Ta-kung-pao Press, October, 1963.
Perkins, Dwight, H., 1966. *Market Control and Planning in Communist China*, Cambridge: Harvard University Press, 1966.
Perkins, Dwight H., 1967. "Economic Growth in China and Cultural Revolution, (1960–April 1967)," *The China Quarterly*, No. 30, April–June, 1967.
Perkins, Dwight H., 1968. "Industrial Planning and Management," in Eckstein et al., *Economic Trends in Communist China*, Chicago: Aldine Publishing Co., 1968, pp. 597–635.
Perkins, Dwight H., 1969. *Agricultural Development in China*, Chicago: Aldine Publishing Co., 1969.
Planned Economy (Chi-hua Ching-chi), Peking.
Price, Robert L., 1967. "International Trade of Communist China, 1960–65," in Joint Economic Committee, *An Economic Profile of Mainland China*, Washington, D.C.: U.S. Government Printing Office, 1967.
Priestly, K. E., 1963. *Workers in China*, Hong Kong, 1963.
Quarterly Economic Review of China, North Korea and Hong Kong, London.
Remer, C. F., 1933. *Foreign Investments in China*, New York: The Macmillan Co., 1933.
Schran, Peter, 1961. *The Structure of Income in Communist China*, unpublished Ph.D. thesis, Berkeley, 1961.

Schran, Peter, 1964a. "Handicrafts in Communist China," in C. M. Li, *Industrial Development in Communist China,* New York: Frederick A. Praeger, 1964.

Schran, Peter, 1964b. "Unity and Diversity of Russian and Chinese Industrial Wage Policies," *The Journal of Asian Studies,* February, 1964.

Seventh All-China Congress of Trade Unions, 1953. Peking: Foreign Languages Press, 1953.

Shen, T. H., 1951. *Agricultural Resources of China,* Ithaca: Cornell University Press, 1951.

Stanford Research Institute, 1964. *The Economic Potential of Communist China,* Menlo Park, Cal., 1964.

State Planning Commission, 1956. People's Republic of China, *The First Five-Year Plan for the Development of the National Economy,* Peking, 1956.

State Statistical Bureau, People's Republic of China, 1956. "Statistical Materials on Agricultural Cooperativization and Distribution of the Product in the Cooperatives during 1955," *New China Semi-Monthly,* 94, No. 20, 1956, pp. 63–65.

State Statistical Bureau, People's Republic of China, *Ten Great Years,* Peking: Foreign Languages Press, 1960.

Statistical Abstract of the Indian Union, New Delhi, 1965.

Statistical Work (Tung-chi Kung-tso), Peking.

Statistical Work Bulletin (Tung-chi Kung-tso Tung-hisn), Peking.

Study (Hsueh-hsi), Peking.

Sun Yu-tang (ed.), 1957. *Materials on the History of Modern Industry in China,* Peking: Science Press, 1957.

T'an Chen-lin, 1959. "Modern Large-Scale Construction in Agriculture Has Begun," *People's Daily,* October 29, 1959, p. 2.

T'an Chen-lin, 1960. "Struggle for the Advanced Realization of the Agricultural Development Program," *People's Daily,* April 7, 1960.

Tao, L. K., 1931. *The Standard of Living Among Chinese Workers,* Shanghai, 1931.

Tawney, R. H., 1964. *Land and Labor in China,* New York: Octagon Books, Inc., 1964.

Teng Chieh, 1966. "Fully Develop the Handicraft Work for Serving Agriculture Better," *People's Daily,* January 4, 1966, p. 5.

Teng Yun-t'e, 1958. *History of Natural Disaster Relief in China,* Peking, 1958.

Trade Unions in People's China, 1956, Peking: Foreign Languages Press, 1956.

Tsen Wen-chin, 1958. *Socialist Industrialization of China,* Peking: People's Press, 1958.

United Nations, 1966a. *Demographic Yearbook,* New York, 1966.

United Nations, 1966b. *Statistical Yearbook,* New York, 1966.

U.S. Bureau of the Census, 1964. *International Population Reports,* Series P-91, No. 13, Washington, D.C.

U.S. Consulate General, Hong Kong, 1965. *Current Background,* No. 753, Feb., 25, 1965.

U.S. Consulate General, Hong Kong, 1966. *Selections from China Mainland Publications,* No. 542, Sept. 19, 1966.

U.S. Consulate General, Hong Kong, 1967. *Selections from China Mainland Publications,* No. 572, April 17, 1967.

U.S.S.R. Central Statistical Office, 1957a. *The Achievements of the Soviet Union During Forty Years in Figures,* Moscow, 1957.

U.S.S.R. Central Statistical Office, 1957b. *The Achievements of the Soviet Union During Thirty Years,* Moscow, 1957.

Walker, Kenneth R., 1965. *Planning in Chinese Agriculture,* Frank Cass and Co., Ltd., 1965.

Walker, Kenneth R., 1968. "Organization for Agricultural Production," in Eckstein et al., *Economic Trends in Communist China,* Chicago: Aldine Publishing Co., 1968, pp. 397–458.

Wang, F. S., 1947. "An Estimate of Industrial Capital, Employment and Production in North China during the War," *Central Bank Monthly,* 2, No. 2, December, 1947.

Wang Hsin-fu and Han Chien-chao, 1957. "Energetically Develop Chemical Fertilizer Industry," *Economic Research,* No. 9, October 9, 1957, pp. 11–15.

Wang, K. P., "Mineral Wealth and Industrial Power," *Mining Engineering,* August, 1960.

Wang Shi-ta, 1935. "A New Estimate of Recent Chinese Population," *Quarterly Review of Social Sciences,* 6, No. 2, June, 1935, pp. 191–266.
Wang Shu-chun, 1965. "Place Irrigation to an Important Position in Water-Conservancy Construction," *Economic Research,* No. 3, March 20, 1965, pp. 40–46.
Wang Wen-ting and P'eng Wei-ts'ai, 1957. "Reform of the Taxation System Has to Be Taken in Two Steps," *Impartial Daily,* August 11, 1957.
Wen-hui Daily (Wen-hui Pao), Hong Kong.
WHO (World Health Organization), *Statistical Yearbook,* 1959.
Willcox, Walter F., 1940. "The Population of China and Its Modern Increase," *Studies in American Demography,* Ithaca: Cornell University Press, 1940.
Wilson, Dick, 1964. "Interview with Chen Ming," *Far Eastern Economic Review,* May 14, 1964.
Wilson, Dick, 1965. "Peking's Trading Plans," *Far Eastern Economic Review,* May 20, 1965, pp. 352–354.
Winfield, Gerald F., 1950. *China: The Land and the People,* revised edition, New York: William Sloane Associates, Inc., 1950.
Worker Daily (Kung-jen Jih-pao), Peking.
Wu Wen-hui, 1941. "An Inquiry into the Land Problem of China," *New Social Science Quarterly,* 1, No. 4, 1941.
Wu Wen-hui, 1944. *The Land Problem and Policy in China,* Chungking: Commercial Press, 1944.
Wu, Yuan-li, 1963. *Economic Development and the Use of Energy Resources in Communist China,* New York: Frederick A. Praeger, 1963.
Wu, Yuan-li, 1965. *The Economy of Communist China,* New York: Frederick A. Praeger, 1965.
Wu, Yuan-li, 1967. "Planning, Management, and Economic Development in Communist China," in Joint Economic Committee, *An Economic Profile of Mainland China,* Washington, D.C.: U.S. Government Printing Office, 1967.
Wu, Yuan-li, Francis P. Hoeber, and Mabel M. Rockwell, 1963. *The Economic Potential of Communist China,* Menlo Park, California: Stanford Research Institute, 1963.
Yang Chien-pai, 1956. *Achievements in the Recovery and Development of the National Economy of the People's Republic of China,* Peking: Statistical Publishing House, 1956.
Yang, Simon and L. K. Tao, 1931. *A Study of the Standard of Living of Working Families in Shanghai,* Peiping, 1931.
Yeh, K. C., "Soviet and Communist Chinese Industrialization Strategies," in Donald W. Treadgold (ed.) *Soviet and Chinese Communism,* Seattle: University of Washington Press, 1967.
Yeh, K. C., 1968. "Capital Formation," in Eckstein et al., *Economic Trends in Communist China,* Chicago: Aldine Publishing Co., 1968, pp. 509–48.
Yen Chung-p'ing, 1955. *A Draft History of Cotton Textiles in China,* Peking, 1955.
Yi Tseng, 1966. "Chemical Fertilizer Industry in Communist China," *Studies on Chinese Communism,* 9, No. 5, May 31, 1966, pp. 53–70.

Index

Africa, 44, 139, 199, 213
Agricultural Bank of China, 125
Agriculture, 2–10, 86, 87–126, 146, 184, 216–217, 222, 223, 224
    capital in, 5, 9, 46, 87, 120, 121, 122
    collectivization of, 34, 36, 37, 39, 41–42, 91, 97, 105–106, 119, 124, 129, 144, 149, 150, 151, 165
        see also Collectives
    development, 12, 34, 48, 87, 92, 104–124
    and economic growth, 88–97
    implements in, 97, 109, 123, 124, 125, 148, 149, 150, 227
    income in, 5, 7, 8, 9, 89, 152, 154
    and industry, 87, 88, 95–96, 98, 125, 153
    investment in, 7, 9–10, 34, 38–40, 48, 92, 106, 121, 125, 135, 152, 183, 215, 225
    labor force in, *see* Labor force, agricultural
    mechanization of, 88, 101, 109, 118, 119, 122, 123
    multiple cropping, 8, 11, 106, 107, 109
    output (productivity), 7, 10, 33, 41–42, 47, 87–95, 97–107, 109, 113, 118, 119, 123, 125, 147, 151, 152, 153, 156, 159, 181, 190, 202, 217, 219, 225, 226, 227
    planning, 93, 151, 158, 159, 161, 163, 164–165
    policy, 88, 104–126, 152
    prices in, 4, 156, 162
    private enterprise in, 148, 150, 151, 154
    products, 5, 8, 99, 104, 107, 108, 124–125, 164, 167, 200, 202, 203, 206, 209
        see also Barley; Cereals; Corn; Cotton; Fruits; Grain; Millet; Oilseeds; Peanuts; Potatoes; Rapeseeds; Rice; Soybeans; Sugar; Tea; Vegetables; Wheat
    raw materials of, for industry, *see* Industry, raw materials for
    reform of, 87–88, 89, 90, 123, 124
    share of, in domestic product, 92
    share of, in net national product, 2
    share of, in total product, 90
    subsidiary activities in, 97, 103–104, 125, 148, 165, 183
    taxes in, 4, 8–9, 87, 149, 152, 156, 157
    yields in, 218–219
    see also Farms and farmers
Aird, John S., on population, 127–129, 131
Albania, 207, 212–213
Algeria, 213
All-China Federation of Trade Unions, 187, 191, 194, 195
Anhwei Province, 26, 74, 112, 136
Animal raising, *see* Livestock
Anshan, 52, 54, 59, 85, 120
Argentina, 100, 208–209
Armament industry, 16, 22
Army, 103, 181
Ashbrook, Arthur G., on population, 128
Australia, 100, 108, 129, 208–209
Autarky, 93, 143, 228

Index

Automobiles, 52, 53, 85, 228
Aviation, 17, 82, 106, 144

Balance of payments, 48, 210–212
Bangkok, 9
Banks, 10, 12, 125, 133, 144, 151, 158, 174
Barley, 6, 100, 198
Begging, 137
Bergson, Abram, on investment, 39
Birth control, 129, 130, 131, 227
Birth rate, see Population, birth rate
Blast furnaces, 53, 54, 59, 77
   backyard, 47
Bombay, 167
Bonds, government, 156–157
Brazil, 6
Brick production, 13
Bridges, 80, 121
Buck, John L.
   on agriculture, 3, 5, 8, 9, 11, 93, 166
   on population, 26, 28, 29, 30, 31
Budgets
   family, 174, 176, 178, 179
   state, 155, 156, 161, 221
Bukharin, Nikolai, 33–34
Burma, 108, 209, 213
Businessmen, 30, 157

Cadres, 123, 144, 150, 153, 159, 165, 196
Cambodia, 213
Canada, 100, 105, 108, 208–209
Canals, 5, 47, 81, 111, 112
Canteens, 187, 190
Canton, 18, 21, 59, 119
Capital, 14, 16, 17, 27, 29, 34, 49, 67, 72, 84, 86, 112, 161
   accumulation, 9, 34, 143, 157
   foreign, 20, 21, 143
   formation, 1, 5, 10, 12, 20, 50, 143, 224
   goods, 61, 88, 157, 226
      import of, 51, 53, 198, 199, 202, 203, 209, 211, 212
   private, 145, 146, 225
   stock, 149
Capital-intensive activities, 43, 44, 137, 228
Capital-output ratios, 70
Cement, 25, 35, 48, 58, 62, 64, 66, 144, 162, 220
Census, 1953, 128, 129, 130
Central China, 18, 73, 74, 77, 80
Central Committee of the Chinese Communist Party, 111, 123
Cereals, 175, 205
Ceylon, 6, 209, 213
Changchun, 52, 53, 85, 120
Changsha, 9
Chao, Kang
   on industrial output, 55, 56, 61
   on transportation, 219

Chao I-wen, on industrial production, 218
Charcoal manufacturing, 11
Chekiang Province, 3, 74
Chemicals, 13, 17, 18, 24, 48, 53, 59, 61, 74, 83, 85, 131, 132, 203, 204
   see also Fertilizers, chemical; Pesticides
Chen, Nai-Ruenn
   on industrial output, 218
   on transportation, 219
Cheng, Y. K., on foreign trade, 219
Chengchow, 53
Chengtu, 80
Chi-chi-haerh, 73
China Committee for the Promotion of International Trade, 99
Chinese Academy of Sciences, 138
Ch'ing Dynasty, 3, 15, 26
Chou En-lai, 43, 106, 121
Chungking, 80
Cigarettes, 25, 63
Civil war, 19, 59, 129, 150, 168, 227
   see also Cultural Revolution
Clark, Colin, on agricultural output, 6
Clothing, 13, 51, 167, 168, 170, 175, 176, 177, 179, 180, 182, 183, 205, 228
Coal, 11, 16, 17, 18, 20, 21, 34, 35, 48, 53, 58, 61, 62, 64, 66, 68, 75, 76, 79, 85, 132, 144, 162, 167, 203, 209, 218
Collectives, 152
   production teams, 150, 151, 153, 154, 159, 195, 196
   work-points, 41, 149, 150, 195, 196
Commerce, see Trade
Commodities, 36, 57, 61, 65, 68, 156, 158, 162, 163, 164–165, 173, 174, 179, 199, 228
   export of, 200, 209
   hoarding of, 17, 19
Communes, 36, 103, 104, 105, 106, 112, 121, 123, 124, 125, 151, 152, 153, 154, 157, 158, 165, 196
Communications, 8, 29, 34, 138, 187, 223
Compradores, 20
Conservation
   soil, 48, 109, 125
   water, 43, 83, 109, 110, 111, 112, 125, 133
Construction, 44, 50, 51, 53, 54, 70, 82–84, 152, 182, 187
   capital, 46, 48, 78, 124, 125, 137, 158, 159
   employment in, 43, 133, 134
Consumer goods, 16, 20, 22, 25, 44, 46, 48, 60, 61, 67, 69, 95, 96, 131, 162, 163, 177, 178, 182, 195, 203, 216
   investment in, 183
   share in domestic product, 38–39
Consumption, 46, 48, 58, 76, 130, 157, 222
   capital, 92
   of food products, 31, 95, 97, 99, 100, 117, 170, 175, 176, 180, 184, 186

Index

personal, 167, 169, 171, 173, 174, 177, 182, 183, 187
Coolies, 10, 59, 167, 193
Cooperatives, 40, 41, 147, 152
  agricultural, 97, 146, 148–150, 157, 158, 165
  handicraft, 146, 147, 157
  industrial, 154
  marketing, 146
  producer, 123, 146, 149
Corn, 6, 167
Cottage industries, 71
Cotton, 5, 6, 16, 21, 22, 101, 102, 103, 104, 107, 108, 113, 124, 125, 151, 152, 156, 163, 164, 203, 206, 215, 219
  cloth, 13, 17, 18, 24, 25, 35, 63, 64, 65, 66, 68, 72, 73, 96, 97, 109, 167, 177, 181, 184, 218, 220, 223
  yarn, 11, 17, 18, 24, 61, 62, 64, 72, 73, 96, 97, 144
Credit, 7, 8, 9, 48
Cuba, 207, 212–213
Cultural Revolution, 59, 60, 126, 158, 161, 182, 189, 195, 196, 197, 202, 216, 217, 227, 228
Customs duties, 155, 156
Czechoslovakia, 207

Dairen, 72, 78, 85, 116, 221
Dams, 106, 111, 125
Dawson, Owen L., on grain production, 93–95, 98, 99, 185
Death rate, *see* Population, death rate
Debt repayment, 211–212
Decentralization, 44, 144, 162, 163
Defense material, 83
Department of Industrial Statistics, 20, 21
Deportation, 36
Depreciation reserves, 156
Depression, 23, 33, 86, 130
Directorate General of Budget, Accounts, and Statistics, 5
Doctors, 30, 31, 141, 142, 187, 188
Domestic expenditure, 39
Domestic product, 38, 50, 92, 143, 154, 169, 209, 210, 216, 218, 222, 224
Draught animals, 5, 39, 147, 148, 149, 164
Droughts, 3, 109, 110, 111, 125, 159
Dzungarian basin, 77

Earthenware, 13, 177
East China, 73, 74, 101
East Germany, 207
Eckaus, R. S., on wages, 14
Eckstein, Alexander
  on food production, 99, 185
  on foreign trade, 201, 203, 204
  on industry, 53
  on population, 128
  on Soviet assistance, 221
Economic development, 26, 28, 30, 33–49, 85–86, 92, 109, 130, 137, 138, 143, 154, 162, 200, 202, 203, 211, 217, 223
  balanced growth model, 33–35, 39, 48
  Indian model, 33–45, 48–49
  Soviet model, 33–42
Economic growth, 215–217, 222–223
Economic heritage, 1–32
Economic planning, 143, 144, 158–162, 199, 223–229
Economic reorganization, 144–154
Education, 29–30, 32, 36, 138–140, 142, 151, 168, 188, 189, 190
  foreign, 138–139
  see also USSR, Chinese students in
  see also Schools; Universities and colleges
Eggs and egg products, 20, 24, 203
Egypt, 6, 108, 140, 141
Electric power, 17, 18, 20, 21, 35, 39, 46, 51, 55, 61, 62, 64, 78–79, 117, 132, 218, 220, 223
  see also Hydroelectric plants
Embroidery, 11, 14
Emerson, John Philip, on employment, 133
Employment, *see* Labor force
Enamelware, 11, 20
Engineers, 30, 53, 84, 138, 139, 140, 141, 193
Entrepreneurs
  foreign, 16, 20, 21, 24
  private, 19, 22
  state, 22
Europe
  Eastern, 54, 198, 206, 207, 212
  Southern, 44
  Western, 51, 228
Expenditures, household, 167, 168, 174, 175, 183, 184
Exports, 2, 10, 16, 23, 24, 87, 88, 95, 97, 101, 104, 163, 177, 198, 199–203, 205–206, 208–212, 216, 225, 226, 228

Factories, 12, 13, 14, 15, 16, 17, 20, 21, 25, 29, 37, 39, 40, 42, 44, 46, 50, 55, 61, 65, 67, 68, 70, 72, 86, 95, 96, 133, 144, 145, 177, 182, 186, 189, 190, 193, 228
Famine, 4, 128, 131, 170, 225
Farms and farmers, 4–10, 11, 14, 27, 31, 33, 38, 93, 121, 124
  see also Agriculture; Peasants
Fengman station, 78
Fertilizers:
  chemical, 5, 39, 41, 48, 51, 61, 62, 64, 65, 66, 71, 98, 99, 100, 102, 109, 112–116, 117, 118, 119, 185, 203, 204, 209, 216, 217, 218, 220, 223, 226, 227
  organic, 97, 109, 113, 116, 117, 118

Field, Robert, on industrial output, 55, 56, 57–58, 59, 60
First Five-Year Plan, 35, 38, 39, 42, 43–44, 50, 51, 52, 58, 61, 67, 70, 72, 73, 75, 76, 77, 78, 80, 82, 84, 85, 90, 92, 93, 95, 97, 98, 100, 103, 104, 105, 106, 110, 111, 116, 117, 120, 131, 132, 137, 151, 154, 156, 157, 158, 159, 160, 162, 164, 183, 199, 200, 215, 216, 220, 224, 225
Fishing, 104, 133
Five-Antis Campaign against private enterprise, 145, 202
Floods, 3, 109, 122, 128, 159
   control of, 152
Flour, 20, 24, 63, 68, 72, 104
Food, 41, 88, 93, 100, 101, 144, 148, 152, 166, 167, 168, 174, 179, 196, 226
   consumption, *see* Consumption, of food products
   distribution, 169
   export of, 205, 208
   import of, 42, 203, 204, 206, 208, 209
   processing, 13, 51, 58, 73, 74, 75, 132
   shortages, 36, 87, 130, 153, 181, 183, 185, 198
      see also Famine
   see also Agriculture, products
Food and Agricultural Organization, 44, 186
Foreign aid, 210, 211, 212–214
Foreign enterprises, 145
Foreign investment, *see* Investment, foreign
Foreign trade, 1, 22–25, 51, 78, 157, 158, 198–214, 215, 229
   balance, 210–212
   composition of, 24, 203–206, 216
   credit accounts, 202, 210, 211, 212
   direction of, 206–210
   with East Europe, 198, 206, 207, 212
   embargoes, 206
   exchange rates, 200, 208, 211
   organization of, 199–200
   policy, 199, 208
   trends in, 200–203
   with USSR, 54, 198, 200, 202, 206, 207, 208, 210, 211, 212, 222
   volume, 23, 24, 25, 200–202, 209, 216, 219, 226, 228
   with West, 198, 203, 206, 207, 208, 209, 216, 226, 228
   see also Exports; Imports
Forestation, 48, 125
France, 138, 206, 208, 209
Freight, 62, 80, 81, 210
Fruits, 176, 205
Fuel, 18, 72, 75–78, 104, 121, 167, 168, 170
Fukien Province, 105
Fur industry, 25, 177

Germany, 206
   see also East Germany; West Germany
Ghana, 213
Government
   Communist, 33, 38, 47, 50, 52, 54, 75, 77, 88, 91, 106, 111, 113, 116, 117, 119, 123, 126, 129, 130, 133, 143, 144, 149, 150, 153, 154, 158–160, 163, 193, 203, 206, 216, 217
   local, 155, 157, 159–160, 163, 195
   Nationalist, 2, 22, 30, 144, 225
Grain, 36, 90, 93, 94–95, 98, 99, 100, 101, 105, 107, 108, 109, 110, 113, 151, 152, 153, 156, 163, 164, 166, 168, 170, 171, 172, 174, 175, 176, 181, 184, 185, 186, 200, 215, 217, 218, 219, 220, 224, 225, 226
   collection of, 41
   import of, 198, 211, 212, 216
Great Britain, 22, 86, 206, 208, 209
Great Leap Forward, 43–47, 49, 50, 51, 57, 58, 60, 61, 65, 68, 69, 70, 71, 72, 73, 76, 78, 79, 83, 86, 97–104, 105, 110, 111, 118, 119, 122, 123, 124, 129, 130, 132, 133, 135, 137, 139, 152, 153, 154, 157, 158, 159, 161, 165, 171, 181, 182, 184, 196, 202, 203, 211, 216, 217
Great Proletarian Cultural Revolution, *see* Cultural Revolution
Gross national product, 2, 34, 39, 155, 214

Handicrafts, 10–15, 131, 132, 133, 134, 168, 173, 203, 228
   capital in, 11, 69
   investment in, 41, 69
   labor force in, *see* Labor force, in handicrafts
   output, 10, 13, 40, 56, 65–69, 95, 96, 104
   share of, in net national product, 10
   wages in, 11, 14, 15
   workshops, 55, 65, 67, 68, 69, 123
Hangchow, 18, 120
Harbin, 52, 59, 85
Harvests, 95, 96, 97, 99, 103, 152, 153, 169, 170, 181, 196
Health conditions, 29–32, 167, 181, 187, 188, 189, 190
   see also Medical care *and* Public health
Heilungkiang Province, 74, 105
Heilungkiang River, 112
Highways, 17, 46, 48, 79, 81, 84, 106, 121, 144, 219
Hirschman, Albert O., on industry, 14, 21
Ho, Ping-ti, on population, 25–26
Hog raising, 97, 103, 104, 164, 165
Honan Province, 22
Hong Kong, 47, 98, 99, 100, 101, 102, 103, 185, 202, 206, 208, 209

Index

Hopei Province, 5, 21, 22, 30
Hou, Chi-ming
  on employment, 45, 137, 218–219
  on industry, 13
  on population, 128, 218–219
Housing, 32, 51, 157, 167, 168, 170, 171, 178, 179, 182, 183, 184, 185, 187, 190
  see also Rents
Hu Shih, 26, 27
Huainan, 76
Hunan Province, 9, 21, 74, 167, 177
Hundred Flowers period, 183, 192
Hungary, 207, 212–213
Hupeh Province, 21, 73, 74
Hydroelectric plants, 53, 78, 79, 85, 112
Hyper-inflation, 1, 19

Imports, 23, 24, 25, 77, 100, 101, 114, 116, 170, 185, 186, 198, 199–204, 208–212, 216, 222, 225, 228
Income, 31, 147, 149, 156, 157, 161, 173, 174, 177, 179, 181, 183, 184, 186, 187, 190, 195, 196, 197
  per capita, 1, 38, 130, 154
India, 3, 6, 7, 30–31, 57, 60, 61, 67, 69, 78, 93, 106, 108, 129, 131, 135, 140, 141, 186, 187–188, 217–223, 228
  five-year plans, 34–38, 41–45, 48–49, 50, 58, 71, 220, 221, 223
  see also Economic development, Indian model
Indonesia, 208–209, 212
Industrialization, 24, 34–40, 43, 51, 52, 55, 57, 77, 85–86, 87, 88, 93, 131, 132, 148, 151, 152, 162, 198, 199, 203, 216, 222, 224, 225, 226, 228
Industry, 15–22, 24, 48, 68, 106, 151, 156, 162, 184, 189, 190, 217
  capital in, 14, 20, 22, 44, 48
  costs, 14, 71, 158, 161
  development, 17, 20, 24, 50–86, 144, 222, 227
  equipment in, 19, 24, 72, 156
  government-owned, 17, 19, 144, 145
  growth, 16, 19, 54–60, 95–96, 183, 203, 216, 223, 224, 225
    pattern of, 60–65
  heavy, 39, 44, 48, 51, 58, 61, 65, 69, 72, 73, 76, 85, 131, 152, 194, 215, 216, 224, 227, 228, 229
  investment in, 20, 21, 34, 38–41, 46, 48, 51, 52, 58, 73, 75, 80, 81, 86, 92, 97, 135, 152, 158, 162, 183, 224, 225
  labor force in, *see* Labor force, industrial
  light, 48, 72, 73, 194
  location of, 72–75
  management of, 44, 194–195
  output, 14, 16, 17, 18, 20, 22, 43, 48, 50, 55–65, 66, 67, 68, 71, 72, 86, 158, 162, 179, 195, 215, 216, 217, 218, 219
  planning, 19, 72, 158, 159, 161–163
  private enterprise in, 145, 146
  raw materials for, 33, 36, 41, 72, 73, 87, 88, 95–96, 137, 145, 148, 163, 208, 209, 214, 216
  share of, in domestic product, 50
  taxes in, 16, 25, 155, 156, 157
Infant mortality rate, *see* Population, infant mortality rate
Inflation, 17, 19, 23, 33
Inner Mongolia, 52, 73
Insecticides, *see* Pesticides
Insurance, 133, 158, 187, 210
  social, 190, 191, 192
Interest rates, 7, 8, 12, 146, 213
International Labour Office, 44, 137
Investment, 34, 35, 42, 46, 111, 144, 156, 158, 168, 173, 180, 182, 202, 222
  capital, 2, 39, 45, 154, 157
  foreign, 15–20, 22, 23, 72
  rate of, 38, 39, 143, 154
  resources, 143
  state fixed, 154–155
  see also Agriculture, investment in; Handicrafts, investment in; Industry, investment in
Iron, 17, 39, 47, 51, 52, 61, 74, 123, 132
  ore, 20, 21, 76, 77, 85, 209
  pig iron, 18, 20, 21, 52, 62, 64, 73, 85, 131–132, 144, 203, 209
Irrigation, 5, 7, 36, 39, 41, 47, 48, 83, 90, 98, 99, 102, 109, 110, 111, 112, 118, 119, 121, 122, 125, 152, 153, 217
Italy, 78, 208–209
Ivory, 11, 14

Jade, 11
Japan, 3, 6, 16, 17, 22, 23, 25, 30, 31, 106, 108, 117, 118, 131, 138, 166, 197, 228
  occupation by, 18–19, 52, 72, 85
  trade with, 199, 206, 208, 209
  see also Sino-Japanese War
Java, 6
Joint enterprises, *see* State-private enterprises
Jones, Edwin F.
  on cotton output, 101–102
  on grain output, 98–99
  on population, 128, 129, 218
  on sown acreage, 107
Jute, 108, 151

Kaifeng, 119
Kansu Province, 77, 175
Karamai, 77

Khrushchev, Nikita, 54
Kiangsi Province, 9, 40, 74
Kiangsu Province, 21, 30, 74
Kiangyin, 30
Kindergartens, 151, 190, 192
Kirin Province, 74, 85, 174, 175, 177, 178
Klatt, Werner, on grain output, 98, 99, 185
Korean War, 33, 145, 162, 221
Kuikiang, 9
Kuznets, Simon, on agriculture, 88, 92
Kwangsi Province, 75, 105
Kwangtung Province, 5, 47, 75, 105
Kweichow Province, 3, 75

Labor force, 10, 27, 34, 37, 45, 49, 73, 179
    agricultural, 28–30, 46, 47, 87, 92, 103, 118, 121, 124, 125, 131, 134–136, 138, 149, 153, 186, 195–197, 216
    educational level, 137–142
    in handicrafts, 11, 40, 67, 69, 167, 191
    incentives, 47, 193, 195
    industrial (nonagricultural), 14, 17, 21, 28–29, 42–44, 70, 131–134, 135, 137, 138, 140, 141, 158, 173, 174, 177, 178, 183, 186, 187, 188, 191, 193, 194, 195, 215–216, 218
    mass projects, 137
    mobilization, 46, 72, 81, 84, 152
    organization of, 191–195
    productivity, 1, 5, 6, 7, 9, 11, 29, 32, 48, 132, 134, 153, 158, 192
    skill, 14, 30, 83, 193, 194
    training, 69, 72, 138–141
    wages, 47, 184
    "workers and employees," 173, 175, 178, 184, 186
Labor-intensive activities, 5, 11, 29, 44, 46, 48, 70, 71, 72, 81, 83, 84, 101, 112, 122, 125, 153, 226, 228
Lanchow, 53, 59, 73, 78
Land, 35
    ownership, 2, 3, 4, 10
    productivity, 6, 7, 105, 106, 107, 217
    reclamation, 48, 106, 152
    redistribution, 93
    reform, 41, 89, 90, 105, 119, 147, 148, 150
Landholding, 91, 106
    concentration of, 5, 9
    fragmentation of, 3, 10
Landlords, 4, 7, 8, 9, 12, 20, 147, 148, 150
Larson, Marion R., on crops, 225
Latin America, 139, 199
Law, study of, 139
Leather, 11, 12, 20, 24, 177
Lewis, W. Arthur, on economic development, 37
Liaoning Province, 72, 73, 74, 85
Libraries, 192
Life expectancy, 30, 31

Literacy, 29, 137, 140, 189, 193, 196
Liu, T. C.
    on agricultural output, 88, 89, 92, 93–94, 95, 166, 218–219
    on capital investment, 38–39
    on clothing, 177
    on domestic product, 218, 222
    on employment, 134–135, 136, 173
    on food production, 185, 218–219
    on industrial output, 55, 56
    on national income, 91
    on population, 26, 28
    on unemployment, 45
Liu Ning-yi, 195
Liu Shao-chi, 147
Livestock, 104, 149, 151, 205
Living standards, 29–32, 86, 166–197
    of peasants, *see* Peasants, living standards of
    pre-Communist, 166–168
    since 1949, 168–171
    urban, 171–183
Loans
    domestic, 155, 156, 161
    foreign, 155, 156, 157
Looms, 20, 21, 25
Loyang, 52, 73, 120
Lumber industry, 13, 55, 62, 64, 66, 74

Machinery, 15, 19, 20, 24, 34, 39, 48, 51, 52, 61, 62, 64, 69, 70, 73, 74, 76, 84, 85, 88, 118, 161, 205
    farm, 48, 119, 121
    imports of, 52, 54, 198, 203, 204, 208, 209, 222
Magnesium, 21
Mah, Feng-Hwa, on foreign trade, 201
Malaya, 208, 209
Malnutrition, 181, 186
Manchuria, 16, 18, 19, 21, 23, 24, 25, 52, 73, 74, 76, 77, 78, 79, 80, 84–85, 116, 119, 174, 182
Manufacturing, 39, 55, 67, 69–72, 88, 132, 137, 145, 146, 210
    see also Industry
Mao Tse-tung, 87–88, 109, 119, 129, 150, 152, 161, 191, 195, 196
Marketing, 8, 9, 21, 48, 69, 72, 88, 89, 103, 228
Markets, 144, 145, 162–165, 179, 226
    free, 153, 165, 174
Marxism, 129–130
Matches, 24, 25, 63
Meat consumption, 166, 175
Medical care, 129, 187
Merchant fleet, 62, 81
Merchants, 8, 147
Mess halls, 151, 152, 153

Index

Metals, 13, 21, 24, 53, 57, 72, 74, 75, 85, 120, 131, 132, 172, 203, 204
   see also Iron; Steel
Metalworking, 34, 39
Middle East, 44, 213
Midwives, 31, 141
Migration, 193, 228
Military conscription, 25, 84
Millet, 6, 167
Minerals, 24, 75–78, 104
   see also Coal; Iron, ore
Mines and mining, 16, 17, 19, 20, 39, 40, 44, 50, 55, 70, 75, 76, 133, 186
Ministry of Economic Affairs, 17
Ministry of Foreign Trade, 199
Ministry of Internal Trade, 199
Modernization, 1, 15–20, 24, 44, 198, 229
Mongolia, 80
   see also Inner Mongolia; Outer Mongolia
Morbidity rate, 30, 31
Moscow, 53
Motors, 81, 209
Mutual aid teams, 40, 148–149

Nanchang, 120
Nanking, 18, 116, 190
National Agricultural Research Bureau, 4, 7, 8, 93
National Economic Construction Bonds, 157
National income, 91
National People's Congress, 160, 199
National product, 2, 10, 46, 130, 168, 169, 170, 212, 215, 220, 221, 222, 223
National Resource Commission, 22
National security, 72
Nationalist government, *see* Government, Nationalist
Natural resources, 85, 220
Nepal, 213
New York Times, 52, 54, 59, 195
New Zealand, 208
North Africa, 44
North China, 4, 5, 6, 18, 29, 31, 80, 101, 109
North Korea, 207, 212–213
North Vietnam, 207, 208, 212–213
Northeast China, 76, 79, 101
Northwest China, 8, 77, 79, 80
Nurkse, Ragnar, 25
Nurseries, 151, 187, 192
Nurses, 31, 141, 188
Nutrition, 31, 225

Oil industry, 51, 53, 59, 72, 77, 78, 80, 131, 218
Oilseeds, 104, 107, 151, 176, 181, 203, 205, 215, 218, 219
Opium, 24
Opium War, 22

Ou Pao-san, on handicrafts, 13
Outer Mongolia, 207, 212–213

Pacific War, 18
Pakistan, 6, 108, 140, 141, 213
Pao-t'ou, 52, 73, 76, 85
Paper manufacturing, 11, 12, 60, 61, 62, 64, 66, 68, 74, 132, 220, 223
Peanuts, 108, 164, 176, 218, 219
Pearl River Delta, 112
Peasants, 2, 3, 4, 8, 9, 11, 34, 41, 42, 48, 87, 91, 103, 106, 110, 111, 112, 119, 120, 123, 125, 147, 152, 154, 159, 164, 165, 226
   households of, 148, 149, 150, 151
   incentives for, 89, 153
   income of, 10, 89, 104
   living standards of, 31–32, 36, 148, 167, 177, 183–186, 196–197
   taxes of, 156, 157
Peddling, 137, 146, 147
Peipking, 166, 167, 179, 180
Peking, 7, 52, 54, 59, 60, 102, 103, 160, 175, 181, 182
Pensions, 187, 189
People's Bank of China, 161
People's Daily, 48, 54, 104, 110, 111, 112, 120, 123, 124, 125
Perkins, Dwight H.
   on consumption, 177
   on grain output, 98, 99, 100, 185, 224–225
   on prices, 173–174
   on wages, 193
Peru, 6
Pesticides, 39, 102, 109, 204, 227
Petroleum, 17, 51, 55, 62, 64, 66, 76, 77, 78, 79, 121, 132, 203, 204, 216
Pharmacists, 141, 142
Philippines, 6, 108
Pi-Shih-Hang Canal, 112
Poland, 207
Population, 1, 2, 7, 25–32, 127–131, 140, 142, 169, 185, 186, 218
   age-sex structure, 128, 129, 140
   birth (fertility) rate, 129, 130
   children, 28
   death (mortality) rate, 30, 31, 129, 130
   females, 128
   growth, 25–26, 36–37, 41, 97, 98, 99, 117, 129, 130, 131, 134, 137, 170, 181, 215–216, 217, 220, 224, 225, 226, 227
   infant mortality rate, 30
   losses, 128
   males, 140, 141
   natural increase rate, 129
   rural, 28, 29, 37
   totals, 129
   urban, 28, 36, 37, 131, 132, 133, 135, 175
Port Arthur, 221

Potatoes, 6, 108, 167, 186
Pottery, 11, 68
Poultry raising, 97, 104, 165
Pravda, 52
Preobrazhenskii, Evgenii, 34
Price, Robert L., on foreign trade, 201, 208, 209, 219
Prices, 162–165, 171, 173–174, 178, 179, 183
  market, 34, 39
Printing industry, 13, 20
Private enterprise, 17, 145, 146, 157, 173
Producer goods, 22, 43, 44, 46, 55, 60, 67, 71, 143, 154, 158, 162, 216
  investment in, 51
  share of, in domestic product, 38–39
Product
  per capita, 169, 171, 220
  per worker, 92
  see also Domestic product; Gross national product; National product
Professions, 137–142
Prospects for Chinese economy, 215–229
Provinces, *see the provinces by name:* Anhwei; Chekiang; Fukien; Heilungkiang; Honan; Hopei; Hunan; Hupeh; Kansu; Kiangsi; Kiangsu; Kirin; Kwangsi; Kwangtung; Kweichow; Liaoning; Shansi; Shantung; Shensi; Sinkiang; Szechwan; Yunnan
Public health, 138, 187, 188
Public utilities, 13, 16, 21, 50

Railways, 16, 17, 19, 33, 35, 59, 60, 79–81, 83, 106, 144, 219, 220, 223
Rangoon, 9
Rapeseeds, 108, 164, 218, 219
Rationing, 84, 144, 176, 179, 224
Raw materials, 12, 14, 21, 22, 24, 48, 68, 104, 117, 118, 146, 198, 199, 205, 206
  see also Industry, raw materials for
Red Guards, 59, 137, 142, 195
Rents, 7, 9, 178, 179, 180, 184, 187
  see also Housing
Reservoirs, 47, 106, 110, 112
Resources
  control and allocation of, 143–165
  markets and prices in, 162–165
  see also Natural resources
Revenue, 155, 221
  domestic, 157
Rice, 6, 9, 68, 101, 104, 108, 117, 167, 198, 226
Roads, *see* Highways
Rubber, 13, 63, 209
Rumania, 53, 207

Saigon, 9
Salt, 63, 64, 133, 209

Sanitation, 188
Sanmen Gorge, 79
Savings, 9, 12, 36, 120, 151, 173, 174, 192, 225
  domestic, 143
  mobilization of, 143, 144, 154–158
  rate of, 154
Scandinavia, 129
Schools, 30, 138–140, 142, 189
Scientists, 30, 137–142
Second Five-Year Plan, 46, 79, 152, 158
Seed, 39, 41, 109, 118, 227
Services, 10, 29, 131, 133, 134, 170, 171, 173, 174, 178, 179, 180, 186, 188, 190, 191
Sesame, 108, 218, 219
Shanghai, 7, 9, 15, 18, 19, 21, 24, 25, 59, 72, 73, 74, 78, 119, 160, 166, 167, 168, 174, 175, 180, 181, 183
Shansi Province, 21
Shantung Province, 5, 21, 25, 119
Shemonoseki, Treaty of, 16, 23
Shensi Province, 22
Shenyang, 85, 120, 182
Shipping, 17, 199
Ships, 20, 80, 81, 228
Shoe making, 11, 51, 104
Sian, 73, 80
Silk, 11, 20, 24, 25, 151
Silver, 11, 14, 23
Singapore, 208, 209
Sinkiang Province, 77, 80, 105
Sino-Japanese War, 5, 16, 17, 19, 20, 22, 23, 26, 168
Sino-Soviet relations, *see* USSR, relations with
Social security, 188
Soda, 20, 62, 64
Soil, 153, 159
  erosion, 106
  fertility, 110, 113
South China, 4, 5, 18, 29, 31, 75, 101, 105, 109
Southwest China, 5, 6, 8, 75, 80, 101
Soviet Turkestan railroad, 80
Soviet Union, *see* USSR
Soybeans, 107, 108, 164, 167, 176, 203, 205, 209, 215, 218, 219
Spindles, 20, 21, 96
Stalin, Josef, 33, 34, 36, 40, 93
Standard of living, *see* Living standards; Peasants, living standards of
Starvation, 181, 185, 186
State Construction Commission, 159
State Council, 160, 199
State Economic Commission, 159, 160, 199
State enterprises, 155, 156
State Planning Commission, 159–160, 199
State-private enterprises, 145–146

State Statistical Bureau, 159, 160
Steel, 17, 18, 20, 21, 34, 35, 39, 42, 48, 51, 52, 58, 59, 61, 62, 64, 65, 66, 72, 73, 74, 76, 77, 85, 86, 120, 123, 132, 144, 162, 209, 218, 220, 223, 226, 228
Strikes, 59
Students, 28, 137, 142, 181
Suanhwa, 25
Sugar, 24, 63, 64, 66, 68, 108, 152, 164, 208, 216, 218
Sulfuric acid, 62, 64, 118
Synthetic fibers, 51, 71, 83, 209
Syria, 213
Szechwan Province, 3, 5, 75, 80, 97, 112

Taching, 56
T'ai-hu Lake, 112
Tailoring, 11, 14, 104
T'ai-p'ing Rebellion, 26, 27, 128
Taiwan, 2, 19, 106, 117, 127, 196, 197, 226, 227
Taiyuan, 119
T'an Chen-lin, 103, 110
Tanzania, 213
Tariffs, 16, 23
Tatung, 76
Tawney, R. H., on handicraft workshops, 12
Taxes, 25, 144, 147, 161, 173, 179
  agricultural, *see* Agriculture, taxes in
  commercial, 155, 156
  excise, 174
  handicraft, 13
  income, 174
  industrial, *see* Industry, taxes in
Tea, 24, 68, 151, 164, 203, 205
Teachers, 139–140, 141, 189
Television, 183
Tenancy, 4–5, 7, 8, 9, 31, 147
Textiles, 15, 16, 20, 21, 22, 24, 25, 42, 51, 72, 73, 74, 75, 96, 97, 132, 181, 184, 206, 228
  export of, 205, 209
  import of, 203, 204
Thailand, 140, 141
Thermal plants, 53, 78
Third Five-Year Plan, 65, 79, 112, 158, 161
Tibet, 160
Tientsin, 18, 21, 24–25, 72, 73, 120, 160
Timber, *see* Lumber
Tinghsien, 30
Tobacco, 16, 20, 24, 108, 203, 205
  see also Cigarettes
Tractors, 52, 83, 119, 120, 121, 122
Trade, 8, 28, 29, 133, 134, 145, 146, 147, 151, 157, 178
  see also Foreign trade
Trade fairs, 103, 153, 165

Trade Union Congress, 192
Trade unions, 187, 191–192, 194–195
Transportation, 8–10, 13, 17, 19, 24, 28, 29, 35, 40, 43, 50, 59, 73, 77, 78, 79–82, 104, 119, 133, 134, 138, 158, 162, 178, 186, 190, 215, 217, 219
  costs, 14, 21
  investment in, 34, 82, 157
  share of, in domestic product, 82
  see also Aviation; Highways; Railways
Treaty of Shemonoseki, 16, 23
Treaty ports, 16, 20, 21, 72
Treaty Powers, 16, 23
Tsin Divide, 112
Tsinan, 18
Tsingtao, 18, 21, 65, 73
Tungting Lake, 112

Unemployment and underemployment, 28, 37, 39, 40, 43–48, 122, 124, 125, 131, 136–137, 215–216, 225
United Arab Republic, 213
United States, 2, 3, 5, 6, 9, 30–31, 45, 77, 79, 82, 105, 108, 129, 138, 206, 213, 221, 228
  Agricultural Officer, Hong Kong, 98, 99, 100, 101, 102, 103, 185
  Bureau of the Census, 128
Universities and colleges, 30, 138, 142
Urumchi, 80
USSR, 30, 168–169, 182, 213
  agriculture in, 3, 5, 6, 41, 93, 95, 105, 108, 121, 150, 152, 196
  aid to China, 51–55, 58, 69, 70, 73, 75, 77, 78–79, 85, 86, 120, 137, 138, 157, 200, 206, 215, 221, 222, 225
  Central Statistical Bureau, 37
  Chinese students in, 138–139, 210
  employment in, 132, 134, 195
  five-year plans, 33, 35–36, 38–40, 50, 58, 67, 95, 132, 170
  industry in, 57, 65, 84, 191, 194, 195
  population of, 37, 129, 130, 131
  relations with, 52, 54, 72, 198, 202, 206, 207, 216, 221, 228
  see also Foreign trade, with USSR
  see also Economic development, Soviet model

Vegetable oils, 24, 63, 68, 200
  see also Oilseeds
Vegetables, 103, 149, 165, 166, 175, 176, 181, 205

Wages, 144, 147, 152, 153, 154, 158, 161, 162, 171, 172, 173, 174, 178, 180, 181, 187, 192, 193, 194, 195–197

Wars, 15, 18, 80, 89, 128
  see also Civil war; Korean War; Sino-Japanese War; T'ai p'ing Rebellion; World War I
Waterways, 219
  see also Canals
Weaving, hand, 11, 15, 25, 68, 73, 96, 104
Welfare benefits, state, 186–191
Wells, 111, 121
West China, 77
West Germany, 51, 208–209
Wheat, 5, 6, 100, 101, 108, 167, 198, 226
Willcox, Water F., on population, 26–27
Workshops, 10, 11, 40, 44, 46, 58, 61
  see also Handicrafts workshops
World War I, 16, 23
Wuching, 59
Wuhan, 18, 52, 59, 73, 76, 85, 120

Yangtze River, 79, 167
  Delta, 112
  Valley, 8, 102, 112
Yeh, K. C.
  on agricultural output, 88, 89, 92, 93–94, 95, 166, 218–219
  on capital investment, 38–39
  on clothing, 177
  on employment, 134–135, 136, 173
  on industrial output, 55, 56
  on national income, 91
  on population, 26, 28
  on unemployment, 45
Yellow River, 53
Yemen, 213
Yenan, 36, 41
Yumen, 77, 78
Yung-li, 116
Yunnan Province, 3, 105